ALCOHOL: The Beloved Enemy

ALCOHOL:

The Beloved Enemy

Dr. Jack Van Impe

with

Roger F. Campbell

THOMAS NELSON PUBLISHERS
Nashville • Camden • New York

Published in Nashville, Tennessee, by Thomas Nelson, Inc. and distributed in Canada by Lawson Falle, Ltd., Cambridge, Ontario.

Printed in the United States of America.

Fourth printing

Library of Congress Cataloging in Publication Data

Van Impe, Jack.
  Alcohol, the beloved enemy.

  Includes index.
    1. Alcoholism. 2. Alcoholism and religion.
I. Campbell, Roger F., joint author.
II. Title.
HV5035.V36   362.2'92   80-16508
ISBN 0-8407-5730-1 pap.
ISBN 0-8407-5192-3 case

CONTENTS

ACKNOWLEDGMENT

We have appreciated the opportunity of reviewing and quoting from Dr. Robert P. Teachout's Th.D. dissertation prepared for Dallas Theological Seminary in 1979. We did not learn of this exhaustive work until near the end of our research and writing. Therefore, quotes from it appear only in the final chapters of this book. Nevertheless, the privilege of reading this masterful work and finding it generally confirming our views has been encouraging.

We also wish to thank Dr. Teachout for his kindness in writing the foreword for *Alcohol: The Beloved Enemy*. Probably no living person has given as much study to the subject of wine in the Old Testament as has Dr. Teachout. His comments about our work and conclusions are gratifying and appreciated.

Dr. Jack Van Impe
Roger F. Campbell

FOREWORD

A century ago there was much material available that the serious Bible student could study in order to decide for himself whether or not God ever approved of the use of intoxicating alcoholic beverages. Many books were written during that period, some of which were by reputable Hebrew and Greek scholars who set forth solid evidence showing that the major words that were translated "wine" in Scripture (Hebrew *yayin*, Greek *oinos*) could mean either fermented or unfermented grape juice. The research of these men clearly demonstrated that the context of divine approval or disapproval in any given passage was the determining factor for interpretation rather than the innate meaning of those words. Because of this availability of literature, preachers were bold to speak powerfully against the use of alcohol by society. Any genuine student of that period of history knows what a beneficial impact the dissemination of this information made upon the United States at the beginning of the twentieth century.

However, since that time the sellers of alcoholic

beverages have done such a thorough job of spreading false propaganda that even conservative Christians have become unsure as to the message of God's own Word on the subject. (Unfortunately Bible scholars have been equally misled by public opinion.) Further complicating this problem is the fact that the earlier literature which could have provided a corrective to this trend has now become virtually unavailable.

When the testimony of the church becomes muted, it is no wonder that alcoholic intake once again becomes unrestrained in society. The effects of alcohol drinking by the majority of this nation's populace during the last forty years have brought this nation down more surely than perhaps any other single factor. Even the secular educators of our country are now becoming more aware of this great peril. (Note especially the recent feature article, "Going Back to the Booze," *Time*, November 5, 1979, p. 71, concerning this alarming trend.)

Therefore it is important both for the purity and power of the church and for the rejuvenation of an ailing America that the long withheld and suppressed message be exposed to all once again: *Alcohol is never approved of by God in any amount for the obedient Christian*. Alcohol is perhaps the most dangerous drug in our culture because of its complete acceptance by society, its total availability to the populace, and its devastating effect on every aspect of our national life—including our homes, our children, our individual health, our safety, our schools, our jobs, our courts, and our government policies.

This book, *Alcohol: The Beloved Enemy*, has been popularly but carefully written to provide the Christian public once again with the information needed to regain an adequate evaluation of alcohol. A major contribution of the book is its research into the real story of prohibition and why it was repealed. This work also documents the devastating damage to the nation (and to the individuals who make up that nation) that alcohol is now exacting every year. It further gives in its final chapter some very helpful suggestions for curbing the abuse of alcohol nationwide. While its treatment of the biblical revelation on this subject is relatively brief, it is comprehensive enough to provide the reader with an overview regarding God's attitude toward the beverages made from the grape.

I recommend this volume to the reader as fully worth his careful study. It is my prayer that the Lord may use it to help in changing the tide of Christian opinion back to a more biblical position on the use of alcohol.

Robert P. Teachout, Th.D.
Associate Professor of Old Testament
  and Semitics
Detroit Baptist Divinity School
Allen Park, Michigan

ALCOHOL: The Beloved Enemy

INTRODUCTION

ALCOHOL: THE BELOVED ENEMY may shock you. The story of alcohol's impact on our nation is chilling . . . frightening. Nevertheless, the story must be told; the facts must be given. When people are informed, they can act responsibly. And public action on America's alcohol problem is long overdue.

During the last century, churches spoke out against the use of beverage alcohol, and many joined in the battle that finally outlawed its sale, for a time. Today, such opposition is fading in many churches. In others, it has long since disappeared. Many former abstainers now imbibe, claiming Christian liberty. What is the proper position on this issue? Does the Bible forbid the use of alcohol as a beverage? Did Jesus make intoxicating wine at the wedding in Cana? Was fermented wine used in the first Communion service? Is social drinking right for some and wrong for others? Can convictions in this area differ and still be biblical? Does Paul's ". . . use a little wine for thy stomach's sake . . ." sanction moderate drinking?

Readers may find the information on the Prohibition Era of special interest. Did prohibition cause increased drinking, as is so commonly believed? Was gangsterism the rule of the day during the "dry years"? Was the Great Depression brought on by prohibition? Careful research has provided some surprising answers.

In our presentation of the problem of alcoholism, we have used the figure of ten million alcoholics and problem drinkers. These numbers are from the Third Special Report to the U.S. Congress on Alcohol and Health from the Secretary of Health, Education, and Welfare, presented in June, 1978. However, a recent news release from the American Medical Association, published in the *Reader's Digest*, sets the figure at seventeen million. This seems to be further proof of the ever-deepening crisis.

Christian response is long overdue.

We will be pleased if many now chained by alcohol's power are set free as a result of this work.

Dr. Jack Van Impe

Roger F. Campbell

1

THE LONG NIGHT

Headlight patterns dance across the bedroom wall, telling of cars passing in the night. But each one is the wrong one. She has watched this all-night movie hundreds of times during her twenty-year marriage.

Momentarily, the fast-moving lines remind her of the brief periods of peace she has known when her man has laid off the bottle. Her attempts to make the good times last have been like trying to capture one of those moving beams on the wall. The light always slips through her fingers, leaving only darkness. She struggles to maintain hope.

Glancing toward the luminous digital clock on the nightstand, she sighs as the numbers change, clunking to a new hour. It is three in the morning. At twelve she could fall asleep hoping for the best. But at three, she awakes to reality.

She's alone.

His place is empty.

The sinking feeling that was under the surface when he walked out the door refuses to be contained.

Questions that have been asked on other long nights

return: *Where is he? How long will he be out this time? Who is he with? Has he been injured or killed? What will he be like when he returns? Why does he keep doing this to me? Why do I put up with it?*

Years ago, she would have been beside herself, weeping uncontrollably. Remembering those breaking experiences, she thinks about the rainbow of emotions she has gone through as the wife of an alcoholic.

Anger.

Frustration.

Despair.

Hope.

Anger again.

She's not sure which reaction is best for her own well-being.

Nothing seems to make much of an impact on him . . . at least not for very long. When the children were younger, she packed a suitcase one long night and left him. *Let him have his drinking friends,* she had thought. *The children and I will make it without him.*

But he had lured her back with promises that seemed sincere. Maybe the toughest part of that temporary separation was her discovery that she still loved him and didn't really want to strike out on her own.

Funny how love can take a beating and endure.

With the sudden changing of the headlight patterns, she is out of bed and at the window.

Flashing lights . . . the police.

But they're not stopping.

Relief.

Strange how the butterflies had come just when she thought she didn't care all that much about what happened to him.

The view at the window holds her.

Where are all these people going at this hour?

How many dark windows shield anxious watchers like herself? Do all her neighbors sleep peacefully? Is hers the only home besieged by booze?

She knows better.

Barry and Madge, just two doors down the street, have been this route. But Madge was different. She decided to fight fire with fire. If Barry could escape into the bottle, so could she. Why miss out on all the excitement of his night life? They began making the rounds together. But Madge got hooked. Soon she had to have a bottle at home to get through the day.

Madge was no "wait-at-home." Now she's waiting it out in a hospital . . . drying out again. And Barry's on the prowl . . . he says Madge embarrassed him too many times.

Lights appear in houses here and there. Early risers. Shift workers. And her worker isn't even home yet.

But he is a worker. She has to give him that. His ambition had been one of the reasons for her attraction to him. And his gentleness . . . there was something tender about this guy that had made her feel safe. She had been sure he would never harm her. And she'd been right. Even liquor hadn't made him mean. She was thankful for that.

Sue, in the next block, had appeared at neighborhood club meetings many times with visible bruises. Her explanations always sounded hollow. Everyone knew Rod had beaten her after another night of drinking.

But something had happened to Rod. After everything else failed, he had decided to "get religious." They say someone at work invited him to church and that he and Sue are going every Sunday. Some say

Rod's a new man. Sue hasn't been wearing any bruises lately . . .

I'll have to talk to her about it. If someone's passing out miracles that work on alcoholics, it's time to get in line.

Getting her man in line would be something else.

Enough of this window gazing . . . better get some sleep. Only a few hours now until rising time.

She moves toward the bed, then hesitates. She's too tense to sleep.

Maybe a snack. Warm milk has relaxed her before.

Making her way quietly down the stairs in the darkness so as not to disturb her sleeping teen-age daughter, she finds her way into the kitchen. How many times has she stood warming milk while few others were stirring?

A glance at the frying-pan clock brings a surprise. Four-thirty . . . and still waiting. Her window gazing had lasted longer than she had imagined.

There's something about a warm cup in one's hand before daybreak that stimulates thinking. *As if I needed any more thinking time. I've had enough solitude to last a lifetime.*

Who would have thought it?

Before they married, he had wanted to be with her all the time. And in those early years they had so much going for them: his good job, their home, a new car, plans for a family, and optimism galore. Where had they slipped off the track?

Was it my fault? Did I drive him to drink? Had I nagged too much, wanted too many things? Had I not been supportive enough? Had my love failed?

Had she overreacted the first time he drank too much? Had her scolding and crying triggered some "get-even" mechanism inside him that now made him unwilling to give up this habit that was draining the life out of their marriage?

I'll have to end this guilt trip. No profit in it.

But how had the problem started?

Beer. First they had enjoyed a few beers together as pick-me-ups and to cool off on hot summer afternoons. Then beer had become a regular item on the weekly grocery list.

"Nobody gets drunk on beer," he had said.

Nobody? She was too bitter to laugh.

Before long, beer belonged. It was present for every evening of television watching; it was part of every cookout; it became the drink to serve when friends dropped in.

When he had begun to climb in the company, there was entertaining to be done. And entertaining meant drinking—wine at meals for gracious dining, sandwiched between cocktail times. "Social drinking."

For some reason, she had sensed danger approaching and had cut down on her drinking. Finally she had quit.

His drinking had increased. Liquor helped him through the rough spots, he said. It allowed him to relax when the pressure was on.

When he wanted to stop drinking, he couldn't.

Some said he was sick. She wasn't sure.

She knew that she was sick of his drinking. Sick of his alibis. Sick of his lies. Sick of his broken promises. Sick of alcohol breath in tender moments—and those were getting few and far between.

Oh, please, no tears. Not now. She had thought that well was dry. What if he came home and found her this way? She had promised herself never to break down in front of him again. A woman has to hold on to some dignity.

And then there's my age. That "Life begins at forty" business doesn't work when alcoholism is involved. What's the use? C'mon now, not the doldrums again. Got to have hope. Depression doesn't help. Have to think it through to find answers.

If only they could start over . . . without booze.

Who manipulates public opinion, creating the illusion that drinking is the inseparable companion of the good life? Hasn't anyone told them about the casualties of that lie?

Lifting her cup, she drinks the last drop of milk. One fist is clenched tightly and resting on the table. She's been through that circle of emotions again: anger, tears, depression, anger.

The sky is beginning to lighten. Maybe that's a good sign.

Her face goes blank at the sound of a tire hitting the chuckhole in the drive. She hears the purr of a familiar motor.

A car door slams. There is a rasping growl of a man clearing his throat; keys jingle; the lock clicks. She watches the doorknob turn.

Which story will he use this time? What will he be like? Apologetic or sarcastic? Stumbling or somewhat in control? Ashamed or arrogant? How long will this go on?

More than ever, she is determined to find the answers needed to enable the two of them to live again.

"The Long Night" is imaginary, yet it is a tale too true. Thousands wait long nights for mates who drink and for others caught in alcohol's cruel trap. With the increase of alcoholism among women, more men are experiencing the agony of witnessing the slow destruction of the ones they love. Often, parents are also members of the waiting brigade.

What are the causes of this growing social problem?

Are there any solutions?

What remedies have been tried in the past?

Is there any hope for alcoholics?

Why is man so susceptible to the ravages of beverage alcohol? Why does he persist in carrying on a love affair with this enemy of body and soul?

Should total abstinence or responsible drinking be our goal?

2

THE LOVE AFFAIR

Entering the front door of his plush suburban home, the well-dressed businessman drops his brief-case at the foot of the stairs that wind elegantly up from the foyer. Graying at the temples, he appears handsomely distinguished and successful. The strain of a heavy day lines his face.

"I'm home," he calls.

Instantly the perfect mate for Mr. Success appears and is lovingly welcomed into his waiting arms.

"You're tired, darling," she sympathizes, gently slipping out of his embrace. "I'll fix you a drink."

"The usual," he sighs, making the most of his fatigue.

Moving easily to the liquor cabinet, she prepares two drinks. Her almost rhythmic movement at the bar reveals that she is going through a familiar ritual. It all seems so natural . . . the thing to do.

Millions of admirers of these "beautiful people" have vicariously lived this homey scene. With a few variations, it will be enacted again and again throughout the year on afternoon soap operas, supposedly portraying the fulfillment of the American dream.

Liquor flows freely on television, as it does in most of the entertainment media. In American drama, alcohol has a starring role: entertaining at social functions of the elite; soothing the nerves of those going through crises; dulling pain for cowboys so that bullets and arrows can be removed; supplying liquid courage to men in battle; lifting the spirits of the depressed, preventing suicides; providing steady hands for doctors, enabling them to perform surgery in the most trying of circumstances; gracing the tables of lovers who rendezvous in dimly lit restaurants, adding romance and intrigue; rewarding heroes who have carried out feats of daring to save lives and protect the innocent.

Alcohol mixes readily with writer's ink. And from Archie Bunker's can of beer to the whiskey flask of the Old West's hired gunman, the message is clear: Booze belongs.

In America, alcohol is big business. More than $30 billion is spent annually by users. But the profit thirst of the liquor sellers is never quenched. The industry spends more than $600 million each year for advertising, employing the latest in sophisticated marketing techniques.

Booze beckons on every hand.

Billboards dot major highways, hawking one brand of alcoholic beverage over others. Sex symbols, entertainers, athletes, and outdoor sportsmen are recruited in advertisers' efforts to get the public to drink more. The strategy, though polished, is elementary: *identification*.

Drink and be successful.

Drink and be desirable.
Drink and be sensuous.
Drink and be lucky in love.
Drink and be athletic.
Drink and be popular.
Drink and be wealthy.
Drink and be a winner.
Drink and feel relaxed.
Drink . . . it's the way to get back to nature.

Is the strategy working? Consider the results.

Since 1939, the American Institute of Public Opinion has polled American adults on their drinking habits by posing the question, "Do you have occasion to use alcoholic beverages such as liquor, wine, or beer, or are you a total abstainer?" The initial Gallup Poll (taken in 1939) revealed that 58 percent were drinkers and 42 percent were abstainers. The most recent poll (1977) shows that 71 percent are drinkers and only 29 percent are abstainers.[1]

A disturbing dimension to the poll was the number of young people who drink. In the age range of 18–29, 78 percent are drinkers.[2]

Interestingly, college graduates seem to be more involved in alcohol use than those with less education. Among this group, 82 percent are drinkers. This may be due to their drive for success, making them more vulnerable to the identification thrust of most liquor advertising.[3]

Whatever the reasons for drinking among both young and old, one thing is certain: Drinking in America is on the increase with annual consumption now at 32.5 gallons of beer, wine, and spirits for every person over fifteen. Alcohol is capturing the hearts of people of all ages and positions in life, laying claim to being one of the ingredients of success and happiness.

But alcohol is an enemy . . . a destroyer.

Historian Sir Arnold Toynbee has identified alcohol as a major force in the destruction of nineteen civilizations preceding our own. And evidence of alcohol's destructiveness is all around us. Nevertheless, our love affair with this enemy continues.[4]

Two people stand before a minister or public official and speak their marriage vows. Later, at a reception, they hold two champagne glasses high, toasting their future. Alcohol has been invited to their wedding. And few things on earth are as likely to destroy their marriage as the one they lift for luck.

An anxious father paces nervously outside a hospital delivery room, awaiting word concerning his wife and their first child.

A nurse appears with good news, and he is on top of the world. Within a few hours, the proud father is on his way to a bar where he will celebrate with friends. A new life has entered the world and the announcement is accompanied by free rounds of the drink that speaks of good times and prosperity. Glasses will be filled again and again as a send-off to the newborn child and as a confirmation of the father's manhood. But alcohol's destructive effect can drive a lifelong wedge between a father and his child.

A group of men prepare for their annual hunting or fishing expedition. The camping gear, sporting equipment, and other provisions are carefully packed into the van . . . including an ample supply of alcohol. How can anyone get back to nature without the natural brew? Alcohol and enjoyment of nature are inseparable companions; six-packs and whiskey bottles are made to blend with trees, mountains, and fast-flowing streams. The advertising team has made this clear. Never mind the fact that some of the men are less likely to return from this trip because booze is going along.

A man relaxes in his favorite easy chair to watch a ball game on television. The can of beer in his hand imparts the feel of the old ball park. As he sips and watches, periodic commercials portraying beer at its cold and foamy best remind him that he's indebted to these benevolent brewers for bringing his favorite sport into his home.

Alcohol and athletics—a strange combination.

Baseball manager Connie Mack has said: "All the umpires put together have not put as many ball players out of the game as 'old man booze.' "[5]

The advertisers have made the wedding of beverage alcohol and ball games seem so right. But the athlete who loves alcohol courts disaster, and television viewers will soon be watching a new hero who has taken his position. In the athlete-and-alcohol affair, to love is to lose.

Alcohol has become a beloved companion of many Americans, sharing their most intimate experiences and being party to decisions that affect our nation and the world.

Bars have become standard furnishings for thousands of homes.

Liquor lubricates the wheels of business, with millions of dollars changing hands while executives linger over drinks.

Cocktail parties have become the most common social gathering for the affluent and are considered important occasions for making business and political contacts.

Holidays are drinking days, when Americans consume huge amounts of alcohol, seemingly admitting they are unable to relax or enjoy recreation without this friendly fluid.

Christmas has become less spiritual and more spirited. Hilarity has replaced meditation on the eternal message of Christ's birth. Office parties during this season are often drinking marathons, degrading those who attend.

Drinking the old year out and the new year in is indicative of the craving for one more fling with alcohol while time remains. Like lovers lingering over a fond farewell, drinking Americans try to squeeze out one last high before the final tick of the clock ends the old year. Then, as if flying to the arms of a new love, they embrace the new year by toasting its arrival.

Taverns and retail liquor stores far outnumber churches in the land. And Billy Sunday's evaluation of the tavern's influence is still timely.

> If all the combined forces of hell should assemble in conclave and with them all the men on earth who hate and despise God, purity and virtue—if all the scum of the earth could mingle with the denizens of hell to try to think of the deadliest institution to home, church, and state, I tell you, sir, the combined forces of hell could not conceive of or bring an institution that could touch the hem of the garment of the tavern to damn the home, mankind, womanhood, business, and every other good thing on earth.[6]

Alcohol is a drink for all seasons. Users depend on it to warm them up, to cool them down, to build their courage, to kill their pain, to pump up their enthusiasm, to calm them down, to chase away their blues, to forget, to enable them to love more fervently or to strike out more viciously.

It is not surprising, then, that the young also fall in love with this product that equips its friends for so many occasions. As a result, the age of those who drink alcohol creeps downward year after year. According to Dr. Ernest Noble, Director of the National Institute on Alcohol and Alcoholism, sixty-two percent of all seventh graders now drink.[7]

A 1977 Gallup Youth Poll showed that the two key reasons why teenagers used alcohol were peer pressure and to escape from the pressures of modern-day life. Other reasons frequently mentioned were "for kicks," problems at home, "showing off," and trying to act "grown up."

A fifteen-year-old ninth-grade girl told the Gallup surveyors, "People my age sometimes follow the group so they won't be outcasts. They try to enjoy themselves, but then things get out of hand."

Some in the Gallup survey cited their reason for alcohol use as "escape from the pressures of being a teenager."

Another high school boy saw "social tensions caused by lack of trust and love in one's family" as a key reason young people turn to alcohol and other drugs.[8]

Once initiated, young drinkers take to the love affair with gusto, deeming alcohol necessary for all important occasions. It becomes a close friend and confidant in times of crisis.

Perhaps alcohol's most dangerous potential for destruction is its power to influence decisions in government. What is the impact of alcohol in the day-by-day legislative, judicial, and executive processes?

How many laws are passed because of politicking at drinking affairs?

How many committee compromises are worked out by key people over cocktails?

How many votes in state legislatures and in Congress are in some way influenced by the alcohol in the blood of elected officials?

How many votes are missed because of hangovers?

How many judicial decisions are tainted by alcohol?

Is a drinking judge competent to hand down decisions affecting the lives of individual Americans for years to come?

Can judges who abuse the product that is the nation's number one drug problem correctly evaluate cases involving other harmful substances?

What part has alcohol played in foreign policy-making during this century?

If .06 percent alcohol in a person's bloodstream (about three drinks) makes a driver twice as likely to cause a traffic accident, what might the same quantity do to a negotiator dealing with problems concerning nuclear weapons?

How much alcohol in the blood would be required to cause a chief executive to make a misjudgment that would bring about a nuclear holocaust?

Long ago, the writer of Proverbs advised:

> It is not for kings, O Lemuel, it is not for kings to drink wine; nor for princes strong drink: Lest they drink, and forget the law, and pervert the judgment of any of the afflicted (Prov. 31:4,5).

Thomas Jefferson wrote:

> The habit of using ardent spirits by men in public office has often produced more injury to the public service, and more trouble to me, than any other cir-

cumstance that has occurred in the internal concerns of the country during my administration. And were I to commence my administration again, with the knowledge which from experience I have acquired, the first question that I would ask with regard to every candidate for office would be, "Is he addicted to the use of ardent spirits?"[9]

If all public officials were called upon to pass Jefferson's test, how many would retain their jobs? How would the love of country fare in contest with this age-old love affair with alcohol?

Man's affection for alcohol spans the centuries. In his book *Almost All You Ever Wanted To Know* About Alcohol *but didn't know who to ask!* Robert L. Hammond writes:

> Ancient warriors danced around their tribal campfires as they drank beer and wine. Because of its peculiar effect on thinking and behavior, alcohol took on a mystical nature. Some regarded it as a magical potion, and the use of wine became common in various religious ceremonies.[10]

The first biblical account of making and drinking wine follows the record of the Genesis flood: "And Noah began to be an husbandman, and he planted a vineyard: And he drank of the wine, and was drunken . . ." (Gen. 9:20,21).

Noah's drinking ended in tragedy. And beverage alcohol is more potent now than it was then. To quote Hammond again:

> The wines and beers of early man were relatively weak by today's standards for alcoholic beverages. Through the process of natural fermentation, when the

beer or wine reached an alcoholic content of about 14 percent alcohol, the yeast cells would die and fermentation ceased.

Although natural fermentation can produce beverages with an alcoholic content up to nearly 15 percent, most beers and wines used by ancient tribes were much milder, usually less than half that strength.

It was not until around A.D. 800 that human ingenuity developed the process of distillation which made stronger alcoholic beverages possible.[11]

Today, civilized people imbibe powerful potions unknown around the campfires of the ancients and then become savages in their homes and on the streets.

We shake our heads and wonder why.

Observing this drama of destruction, Shakespeare lamented:

> Oh! that men should put an enemy into their mouths to steal away their brains! O God, that we should with joy, pleasure, revel, and applause transform ourselves into beasts![12]

Was Shakespeare an alarmist?

Is beverage alcohol really that destructive?

Should we meddle with this longstanding love affair of man?

3

THE DESTROYER

Beverage alcohol kills more than 200,000 Americans every year.

Each decade two million drop from our ranks, the victims of man's old and deadly enemy.

With the passing of each generation (forty years), alcohol has slain eight million in this country—outdoing the efforts of Hitler and his Nazis during the Holocaust in which six million Jews died in Europe's death camps.

At 8:15 A.M. on August 6, 1945, the first atomic bombs used on human beings were dropped on Hiroshima and Nagasaki, Japan, ushering in the apocalyptic nuclear age. Before the war, Hiroshima had been the seventh largest city in Japan with a population of 340,000. When the atomic explosion was over and the nightmarish mushroom cloud had ascended, 80,000 of Hiroshima's citizens were dead. In Nagasaki, 35,000 lives were snuffed out. The misery among the living in both cities remains unthinkable.

Most Americans would rather not dwell on the awful destruction of Hiroshima and Nagasaki; many are still debating whether there might have been another way to end the war. Still, we tolerate two Hiroshimas and a Nagasaki each year in our own land. No wonder Edwin R. Anderson entitled his tract on alcohol: "The 'A' Bomb Has Already Fallen."

The war in Vietnam was one of the most tragic chapters in American history. Apart from the Civil War, no conflict has ever found the nation so divided, so torn apart. Families were at war among themselves over national policy. With the benefit of hindsight, some have changed positions since the war ended, but refugees from that troubled land are still seeking shelter and a place to live in peace.

Casualties in Vietnam were high: More than 57,000 Americans died in the nine years of fighting. But on another front the casualties were higher: During the same period, alcohol killed nearly two million at home.

On December 24, 1965, in Phoenix, Arizona, a drunken man driving a pickup truck sped through two red lights at eighty miles an hour. Seconds later, he veered across the center line and smashed head on into a station wagon.

In the car were a father and mother and two teen-age children on their way from Idaho to San Diego to meet a son returning from Vietnam. The parents were smashed against the instrument panel and died instantly. The two teen-agers were hurled from the car into the street, where they lay critically injured.

The drunken driver in the pickup truck was so entrapped in the twisted wreckage that it took rescuers using a blowtorch thirty-five minutes to extricate him. But that hardly mattered. He had died in the shattering impact.[1]

The soldier son arrived safely in San Diego, having

survived the dangers of battle. But his family had been cut down by a different enemy.

The beloved enemy may not capture as much television coverage or as many headlines as a foreign war, but if you are one who works in the trenches you will see as much blood and misery as on any battlefield. Ask the ambulance drivers, the policemen, the doctors and nurses—even the family members of those addicted to alcohol.

Highway deaths due to alcohol are a continuing national nightmare. The curtain never goes down on this deadly drama. Last year, 49,500 Americans died in 43,500 auto accidents. Alcohol was a factor in more than half of these tragedies.

Often nondrinkers are victims.

"Crash Kills 'Safe Driver' and Five Others," read the headline. The news article told the heartbreaking story.

> A national highway safety leader, his wife and a son died in a two-car crash which took six lives. Another son is in critical condition. On a straight stretch of highway an oncoming car swerved across the center line and slammed into the car coming in the opposite direction. State Patrol Sgt. T. H. Embry said all three of the men in the weaving car had been drinking heavily. The official report said the trio had been refused more beer at two different places shortly before the wreck.[2]

Any other plague claiming 24,000 lives a year would call forth a national effort to root it out and eradicate it. Telethons would be staged to raise money to fight the destroyer. Rewards would be offered to persons finding a cure. Politicians would run on platforms dedicated to conquering the plague.

But beloved alcohol marches that number into the

grave each year, while leaders lift their glasses in toasts to life.

To their credit, some in government are beginning to recognize the serious problems alcohol use has created in our nation. A recent report to Congress minces no words about the destructive role drinking plays in fatal accidents and in social problems. The section of the report entitled "Alcohol-Related Accidents, Crime, and Violence" opens as follows:

Violence, accidental or intentional, constitutes a substantial part of all mortality, illness, and impairment in the United States. Violence plays an especially prominent role in death and injury among younger age groups. For example, accidents are the leading general cause of death for all ages 1 to 38. Research shows that alcohol often plays a major role in such violent events as motor vehicle accidents; home, industrial, and recreational accidents; crime; suicide; and family abuse.

Traffic accidents are the greatest cause of violent death in the United States, and approximately one-third of the ensuing injuries and one-half of the fatalities are alcohol related.[3]

The facts are in. Alcohol is a drug that causes violence, injury, and death.

Still, the general public remains unmoved. Impressive statistics have not proved effective in getting the message across.

Alcohol expert Robert L. Hammond puts the problem in focus with practical illustrations. He writes:

There has been much concern over crime in the streets, and many fear going out after dark because of muggers, yet more people die from alcohol-related car crashes than from homicides. If statistics are an accurate indicator, drunk driving is a far larger problem than criminal homicide. Approximately 18,000 people a year are homicide victims, while some 24,000 per year are killed in alcohol-related car crashes.

To put it another way, more than 400 persons die every week in alcohol-related car crashes. This is about the number of people that can ride aboard a 747 jet airplane. What would be the public outcry if a 747 jet airliner crashed every week killing 400 persons?

In a sense, there is a 747 crash every week, claiming that number of lives in alcohol-related auto crashes on America's highways.[4]

It would be unfair to say that the public is unconcerned about traffic deaths and injuries. We have tried everything from seat belts to reduced speed limits in the interest of highway safety. But we have scarcely dented the accident figure. In spite of all our efforts, automobiles remain unsafe.

Why? Because we refuse to deal with the principle cause of car crashes: drinking drivers.

A common misconception about alcohol's highway killing is that most of it is done by drunken drivers. Actually, even a small amount of alcohol can make an otherwise capable driver a threat to himself and to others.

In his article "What Two Drinks Will Do to Your Driving," Don Wharton claims that drinking drivers are an even greater menace than drunken drivers—because there are so many more of them. He writes:

By focusing attention on "drunken" drivers, who are relatively rare, it whitewashes "drinking" drivers, who are almost numberless. A study of traffic around Evanston, Ill., showed that for every "drunken" driver there were about 30 who had been drinking. A report on

17,000 traffic accidents in Michigan reveals that about three times as many accidents were caused by drivers "who had been drinking" as by those actually "under the influence."[5]

Even a casual acquaintance with the enemy may make you a highway death statistic.

How does a little alcohol affect drivers? Wharton explains:

(1) It slows down reactions. "The average man after one large whiskey," according to New Zealand's Road Code, "will take about 15 percent longer than usual to depress his brake or swing his wheel in an emergency."

(2) It creates a false confidence. New Zealand's Road Code put this neatly: "A little alcohol has the double effect of making him drive worse and believe he is driving better."

(3) It impairs concentration, dulls judgment. Alcohol makes drivers talk more and causes their attention to be more easily diverted.

(4) It affects vision. Dr. Goldberg (of Sweden's Caroline Institute) conducted laboratory tests which showed that moderate drinking caused a 32 percent deterioration in vision. "Alcohol has the same effect on vision," he concluded, "as driving with sunglasses in twilight or darkness; a stronger illumination is needed for distinguishing objects and dimly lit objects will not be distinguished at all; when a person is dazzled by a sharp light it takes a longer time before he can see clearly again." A British ophthalmologist found that alcohol reduced peripheral vision—the capacity to see out of the "corner of the eye" and spot vehicles coming from side roads or pedestrians stepping off curbs.[6]

Drinking drivers see poorly. And some of their victims are not seen at all until it is too late.

It's not a new revelation. Hear Solomon on the subject:

> Who hath woe? who hath sorrow? who hath contentions? who hath babbling? who hath wounds without cause? who hath redness of eyes? They that tarry long at the wine: they that go to seek mixed wine. Look not thou upon the wine when it is red, when it giveth his colour in the cup, when it moveth itself aright. At the last it biteth like a serpent, and stingeth like an adder. Thine eyes shall behold strange women, and thine heart shall utter perverse things (Prov. 23:29–31).

Although traffic fatalities and injuries loom large on alcohol's grisly trail of destruction, that is but one area of the enemy's destruction. Alcohol is a threat to safety in nearly all of life's activities.

Up to eighty-three percent of all fire fatalities and sixty-two percent of all fire burns involve alcohol, as do up to sixty-nine percent of all drownings.[7]

Alcohol is the fatal factor in up to seventy percent of all deaths by falling. Sixty-three percent of those injured in falls have had their balance affected by alcohol use.[8]

According to a recent study, a high percentage of noncommercial aviation crashes may be caused by alcohol. As many as forty-four percent of the pilots involved in accidents during the study had been drinking.[9]

And then there is the frightening factor of alcohol use in the world of industry. One researcher estimates that up to forty-seven percent of nonfatal industrial accidents and up to forty percent of fatal industrial accidents are alcohol related. The report reveals that problem drinkers are as much as three times more likely to be involved in industrial accidents than the general population.[10]

How widespread is alcoholism in industry?

Some sources estimate that one out of every ten industrial employees is an alcoholic. Add to these statistics the many drinkers who are not considered alcoholics but who often drink too much, and you have an immense potential for industrial accidents, to say nothing of the costs involved or the inferiority of the goods produced. People in the know hope they will not purchase an automobile or other item manufactured on a Monday after a weekend binge.

From the highway to the marketplace, alcohol is an enemy, a destroyer. The full impact of its devastation is difficult to grasp. Even America's most famous booze fighter, Billy Sunday, struggled for words to describe alcohol's destruction.

> I wish I could unlock the door, gentlemen, that conceals the secrets of this damnable charnel house, but I am here to tell you, sir, that God never gave any man imagination powerful enough nor lips nor tongue eloquent enough to picture its damnable wreckage and its ruin.
>
> If hydrophobia [rabies] produced one millionth part of the disease and trouble the saloon causes, every dog in America would be killed off before Monday morning.[11]

In reaching for words to expose the enemy that day in 1917 in Pittsburgh, Billy Sunday found one that deserves our study: *disease.*

What does alcohol do to our bodies?

Is beverage alcohol really harmful to health?

4

BOOZE AND YOUR BODY

"I'm paying for what I did during the war," said the man, whose last moments were clearly ebbing away.

The minister standing beside the sufferer's hospital bed was about to protest, thinking that the dying man had taken his illness as a judgment of God for bearing arms for his country. Then, sensing a need for silence, he stifled his impulse to interrupt and listened.

"We were flying bombing missions over Europe," the patient explained, "and drinking whiskey was the only way I could get up courage enough to go. Now my liver has given out." He repeated his lament: "I'm paying for what I did during the war."

Liver damage is only one of the results of drinking, but it is one of the most serious. When alcohol attacks the body, this vital organ is its primary target. Alcoholic liver disease is a serious problem worldwide. Cirrhosis of the liver is the fourth leading cause of death in large U.S. cities for persons aged twenty-five to forty-five.[1]

Although you may have given little thought to your

liver, it is absolutely necessary to your survival and is a wonderful example of our Creator's designing ability. The liver is God's great purifier, continually filtering out harmful chemicals that have entered our bodies. Some of this daily chemical barrage is the result of modern technology and industrial pollution. Even more devastating, however, is the amount of mood-altering chemicals intentionally taken by millions of Americans. Alcohol is one of those most damaging to the liver itself.

Abraham Lincoln once said that alcohol has many defenders but no defense. The defenders of beverage alcohol contended for years that liver damage was not caused by alcohol but by the inadequate diets of those who drink too much. Now that defense has been toppled.

Dr. Charles S. Lieber, chief of the liver disease and nutrition section of Bronx Veterans Administration Hospital and professor of medicine at Mount Sinai School of Medicine, has conducted experiments on baboons showing the destructive effect of alcohol on the liver despite good diets.

Describing Dr. Lieber's experiments, Arthur Fisher wrote:

> Under controlled conditions, a group of baboons was supplied with drink as well as adequate diet for prolonged periods starting in 1968. Of 15 baboons, all got fatty liver, five got alcoholic hepatitis, and another five went on to get cirrhosis—the first time alcoholic cirrhosis had been produced in an experimental animal.[2]

Fisher quotes Dr. Lieber as saying:

> The appearance of the cirrhotic baboon liver was so similar to that of a human cirrhotic liver that the two

were hard to distinguish under the microscope. Now for the first time we can answer yes to the question: Can you produce all aspects of alcoholic liver disease despite an adequate diet?[3]

How much alcohol does one have to use to damage his liver?

Fisher reports that Professor Werner K. Lelbach of the University of Bonn in Germany found liver damage short of cirrhosis, including fatty liver, in male alcoholics down to levels of only eight to twelve ounces of whiskey a day, approximately the amount that a host would serve in four highballs.

Heavy drinking literally bombards the liver, taxing it beyond its capacity to function properly. The liver's ability to handle alcohol, filtering out the chemical impurity, is limited to about one drink in two hours. Since few people linger that long over a drink, the excess enters the bloodstream, carrying its harmful effects to the brain and other vulnerable parts of the body.

In an article, "Booze: Why You Shouldn't Drink a Drop," Toronto *Star* staff writer Sidney Katz, who studied alcoholism at Yale University, writes:

> The persistent social drinker, who imbibes 12 ounces daily for 20 years, has a 50-50 chance of developing cirrhosis of the liver. If a person with cirrhosis continues drinking, chances are 2 to 1 that he'll be dead within a year. Even if you drink as little as 8 ounces of alcohol daily, there's a good chance that you'll develop a pre-cirrhotic state known as "fatty liver." An accumulation of fat builds up because the liver has been taken off its regular job of burning off fat and has to burn away alcohol instead.[4]

Booze, then, is a polluter. Man has been equipped by his Creator with a filtration system capable of handling a certain amount of environmental pollution, but there is a limit to its capacity. Since we live in a time of increased pollution in the environment, it is all the more important that we stop our intake of alcohol, the primary cause of overloading our livers.

The most flagrant, though often unknown, hypocrite of the last half of the twentieth century may be the person who campaigns for filters on every possible polluter of the environment and then uses alcohol as a beverage, impairing his own built-in filtration system. He is an unwitting saboteur of sanitation . . . within his own body. And that can only lead to a poorer quality of life for all.

Alcohol is also damaging to the brain. Because it is so soluble in water and needs no digestion, after overloading the liver's capacity for filtration, alcohol speeds to every organ in the body. It especially affects the brain, acting as an anesthetic and progressively paralyzing the drinker's control centers.

In his *Reader's Digest* article, "Alcohol and Your Brain," Albert Maisel points out the pattern of paralysis experienced by drinkers as a result of alcohol's influence on the brain.

Physiologists have long recognized that many of the familiar effects of drinking are really manifestations of alcohol's effect on our brains. In fact, they have established a direct relationship between the quantity of alcohol we put into our bloodstreams and the area of our brain the alcohol affects. If, for example, a 150-pound man consumes two bottles of beer on an empty stomach, the level of alcohol dissolved in his blood will reach about five hundredths of one percent. At this level, the normal activity of the cortex, or outer layer of the brain—particularly in the centers concerned with worry or anxiety—will be affected. The drinker will feel falsely "lifted up," because the inhibitions that usually hold him steady have, in effect, been paralyzed.

If he drinks enough to raise his blood alcohol level to about ten hundredths of one percent, activity in the motor centers at the back of his brain will be depressed and he'll begin to lose the ability to control his muscles. If his blood alcohol level rises to twenty hundredths of one percent, the deeper portions of his mid-brain will become affected and he'll become increasingly sleepy. Should the level pass one half of one percent, the respiratory centers in the lowest part of his brain may become paralyzed and the drinker will quietly pass from stupor to death.[5]

Frightening? To say the least.

Those silly, cute antics of the slightly tipsy drinker are not signs of "feeling good," but evidences that his brain is being anesthetized.

Alcohol works on the brain much like chloroform, which puts sections of our control and thinking centers to sleep. As the brain becomes less responsive with the increase of alcohol in the bloodstream, the drinker becomes more and more under the drug's control.

At one time it was thought that the brain completely recovered from this forced slumber. Now we know that this is not true. Booze kills these sleeping brain cells. Some of them die with every drink you take. And they are irreplaceable.

Recent studies by Prof. Melvin H. Knisely and his associates, Drs. Herbert A. Moskow and Raymond C. Pennington, at the Medical University of South Carolina have shown that alcohol destroys brain and other body cells through a phenomenon known as "sludging." Alcohol causes groups of red blood cells to stick together so that many capillaries are plugged. As a result, cells in those areas die for lack of oxygen.

People on strict diets or those who are fasting are playing with fire when they use even small amounts of alcohol. Since the brain continually needs sugar, when one is not eating regularly this vital fuel must be produced by the liver. Alcohol disrupts normal liver function, thus halting its production of sugar for the brain. When that happens, permanent brain damage can result.

Children are particularly susceptible to this condition. Dr. John Wilson, assistant professor of pediatrics at Vanderbilt University, warns: "Just one can of beer could affect the intelligence and motor areas of the brain."[6]

Dr. William Altemeier, director of pediatrics service at General Hospital in Nashville, Tennessee, has pointed out that severe irreversible brain damage can result from even a small amount of alcohol consumed by preteen children.

> Alcohol is a special danger to children because it tends to cause hypoglycemia, which is a drop in blood sugar. . . . The brain needs sugar to function, so if the blood sugar drops for long enough, brain damage or retardation can occur.[7]

Children have long been indirect victims of alcohol. Drinking parents often neglect their children, and sometimes they physically abuse them. Who can measure the suffering of little ones whose parents are addicted to this dehumanizing drug? Oceans of tears have been shed by children cowering behind closed doors, listening to drunken parents fight. The emotional trauma that man's beloved enemy brings to the children of the world is mind-boggling. But this is only part of the story.

Now we know that alcohol's attack on the young can begin before birth. Eighteenth-century doctors insisted that alcoholic women were likely to give birth to retarded children, but their theories were ignored. After three hundred years of ignoring these warnings, researchers have discovered them to be valid.

The fetal alcohol syndrome, a condition caused by habitual drinking during pregnancy, is now considered the number one cause of preventable birth defects. The consequences of the fetal alcohol syndrome are especially alarming when one considers that drinking among women is rising steadily. Nearly half of America's ten million alcoholics are women.

A Seattle medical researcher says it is no longer uncommon for an infant to come into the world with the smell of alcohol on its breath. Difficult as it may be to imagine, an increasing number of babies are being born drunk because of their mothers' alcohol intake.

Writer Ronald Kotulak, whose article on the fetal alcohol syndrome appeared in the Chicago *Tribune*, writes that every drink a pregnant woman takes hits her fetus with a chemical sledgehammer. He backs his charge with the following explanation.

> For years society has tended to approve of moderate drinking for expectant mothers despite warnings from older civilizations. It wasn't until the 1970s that the old warning was rediscovered by a University of Washington team.
>
> They lined up eight children of an alcoholic mother and discovered that four of them had the same malformations, which they later called the fetal alcohol syndrome. The defects were striking. The four had small

eyes with short eye slits; shorter noses; smaller mid-face area; flatter facial contours; the line of the upper lip was straight instead of being bowed; and they had smaller heads, indicating smaller brains. The average IQ of children with this syndrome ranges from 60 to 70 and some are as low as 30.[8]

Kotulak says that at first it was assumed that the fetal alcohol syndrome only affected children of heavy drinkers, but that subsequent research showed this is not the case. Five to six drinks a day significantly increases the risk of fetal alcohol syndrome, and mild cases appear when the mother has had not more than two drinks daily.

Dr. James Frias, director of the Birth Defects Center at the University of Florida, says that his research shows heavy drinking ranks second only to German measles in causing environmentally induced birth defects.

As a result of the dangers of the fetal alcohol syndrome, the National Council on Alcoholism has said that it will advise all pregnant women to abstain from drinking. In addition, the Food and Drug Administration has requested that warning labels be placed on all alcoholic beverage containers telling of the dangers of drinking during pregnancy.

Although people do not generally think of alcohol as a drug, that is exactly what it is. And it is the number one drug problem in the United States.

Dr. Marvin Block, chairman of the American Medical Association's Committee on Alcoholism, said:

Ours is a drug-oriented society, largely because of alcohol. Because of its social acceptance, alcohol is rarely thought of as a drug. But a drug it is, in scientific fact.[9]

Former Iowa senator Harold Hughes agrees, saying:

> Many people think alcohol is a stimulant. Others call it a depressant, but it's a drug—dirty, vicious, and brutal. It's the mainstay drug of the American-built society.[10]

Drinking nearly drove Hughes to suicide. But at the breaking point he turned to Christ, put away the intended death weapon, and went on to a productive life. Many do not. Conservative estimates place the blame for one third of all suicides on alcohol. Some recent studies show that the figure may be far higher ... more than sixty percent.[11]

The Department of Health, Education, and Welfare has now linked alcohol use to cancer. Here are the facts from the secretary's report.

> Alcohol is indisputably involved in the causation of cancer, and its consumption is one of the few types of exposure known to increase the risk of cancer at various sites in the human body. In comparison to the general population, heavy consumers of alcohol always show a marked excess of mortality from cancers of the mouth and pharynx, larynx, esophagus, liver, and lung.[12]

Finally, to the list of health problems caused by alcohol, alcoholism itself must be added. The World Health Organization has recently issued a worldwide alcoholism alert, stating that alcoholism now ranks among the world's major health concerns.

A recent item in the *Medical Tribune* reports that in England and Wales, admissions to hospitals for alcoholism have increased some twenty-fold over the past twenty-five years. In Honduras, drinking problems affect sixty-five percent of the rural population, while in Yugoslavia alcoholism was the first diagnosis for half of all male patients admitted to psychiatric hospitals in 1972. Even France, with its "we-can-hold-our-liquor" reputation, spends forty-two percent of its total health budget on the treatment of alcohol-related illnesses. Half of all the hospital beds in that nation are occupied by alcoholics. More than 22,000 Frenchmen die of cirrhosis of the liver each year as a result of steady drinking.[13]

In the United States only cancer, heart disease, and mental illness surpass alcoholism in the number of victims claimed. And it is impossible to determine exactly how much heart disease, cancer, and mental illness is caused by alcohol.

We know that alcoholism is deadly, venting its fury on the entire human body. Canadian Sidney Katz offers the following facts:

> Alcoholics, according to recent research, live 10 to 12 years fewer than other people. Some startling figures came to light when, a few years ago, an analysis was made of 130,000 Ontario alcoholics between 20 and 70 years old. Although they formed only 3 per cent of the population they accounted for these percentages of death: cirrhosis of the liver, 65 percent; peptic ulcer, 23.6 percent; cancer of the larynx, 32 percent; stomach cancer, 19.8 percent; cancer of the mouth and throat, 12.1 percent; poisoning, 53 percent; accidental falls, 25 percent; fire, 48 percent.[14]

Alcohol is harmful to human health. There is no doubt about it. Nevertheless, Americans consume more than five billion gallons of alcoholic beverages annually.

Why do we continue to hurt ourselves?

Why do we tolerate this plague in our land?

What is there about this enemy that captures our affection while injuring our bodies and shortening our lives?

Answers about man's love affair with the destroyer do not come easily. Our search will demand brutal honesty. But even at this point it is clear that Solomon's conclusion about beverage alcohol stands today: "Wine is a mocker, strong drink is raging: and whosoever is deceived thereby is not wise" (Prov. 20:1).

5

PUBLIC ENEMY NUMBER ONE

Crime is a national scandal.

Every thinking American is concerned about crime. Millions are afraid to be on the streets in their own communities after dark. In official recognition of the problem a decade ago, Congress passed the "Omnibus Crime Control and Safe Streets Act," a high-impact program that poured millions of dollars into major cities to fight crime. Within a few years, the program was called a flop.

Money did not buy safety. Crime marches on.

It is difficult to escape the thought of a possible encounter with a criminal. Every branch of the media keeps us aware of crime's danger. Television watchers start the day with some kind of early morning news report that describes all the violence and thievery that took place the night before. Lunch is seasoned with the noon news, and the dinner hour is treated to a full-color reenactment of the crime and terror going on in the world. Viewers are then given an almost steady diet of detective shows and police shootouts until the late news, when they are offered one last opportunity before retiring to get in on the murder and mayhem that somehow might have been missed in the course of the day. It is no wonder that sleeping pills and antacids sell well.[1]

If you think crime is being overcovered by writers and reporters, read what former Attorney General Ramsey Clark writes in his book *Crime in America*.

> Most crime is never reported to police, and much crime is inaccurately reported. Erroneous crime statistics are often used to create the impression that the new chief is doing a good job or to support a movement to add more police. Frequently an apparent increase in crime really reflects an improved effectiveness in law enforcement or in the reporting of crime itself.
>
> The better the police, the more they learn of crimes that are actually committed. Today, some cities suffering far more crime than others take false comfort from lower rates of reported crime because of weaker police departments.
>
> Wherever crime is not reported we know the police are not effective, trusted, or respected. Too many people there would rather suffer crime in silence than report it. It is a bother, useless, gets you involved—or even worse, in trouble yourself.[2]

The crime we know about, then, is but the tip of the iceberg. This fact presents an alarming scenario for the future.

Former ambassador and congresswoman Clare Boothe Luce, writing on crime for the *U.S. News and World Report*, foresaw two awful possibilities in America.

> Assuming that the present growth rate of crime, alcoholism, drug taking, and commercialized sex per-

sists into 1996, America by then will be the most drunken, drug-soaked, sex-ridden, and criminal society on earth.[3]

As to the action the public may take to solve this ever-growing problem, she wrote:

If our Democratic form of government continues for another two decades to fail in the discharge of this responsibility, it is bound to collapse, and our people bound to turn to some other form of government that offers the promise of restoring order. The form of government we will turn to is almost certain to be Fascism, National Socialism, or Communism. These are the modern forms of government associated with the restoration of order to disintegrating and unstable societies. They are also the very systems we fought so long and hard against in the 40s and 50s and precisely because they restored order at the price of justice and liberty.

This is the American tragedy I foresee: Our Democracy, having grown too soft on crime and corruption, will be forced to avoid Anarchy by yielding to a system that will be brutally hard on justice and liberty.[4]

The crime problem is for real. What can we do about it?

Social scientists and criminologists seem baffled by the continued avalanche of crime. Their solutions fail or are only mildly successful.

In his book *Thinking About Crime*, James Wilson questions present theories about crime's causes.

The "Social Science View" of crime is thought by many, especially its critics, to assert that crime is the result of poverty, racial discrimination, and other privations, and that the only morally defensible and substantively efficacious strategy for reducing crime is to attack its "root causes" with programs that end poverty, reduce discrimination, and meliorate privation. . . .

Such a theory, if it is generally held by Social Scientists professionally concerned with crime, ought to be subjected to the closest scrutiny, because what it implies about the nature of man and society are of fundamental significance. Scholars would bear a grave responsibility if, by their theoretical and empirical work, they had supplied public policy with the assumption that men are driven primarily by the objective conditions in which they find themselves. Such a view might be correct, but it would first have to be reconciled with certain obvious objections—for example, that the crime rate in this country is higher than that in many other countries despite the fact that the material well-being of even the lowest stratum of our population is substantially greater than that of a comparable stratum in countries with much lower crime rates, or that crime rates have increased greatly during the very period (1963 to 1970) when there were great advances made in the income, schooling, and housing of almost all segments of society.[5]

Students of the Bible are not surprised when solutions to moral and social problems based entirely on environmental conditions fail. The problem lies deeper. We are socially and morally in trouble because of man's sinful nature. The psalmist David capsulized the root cause of all of the problems of the human race, "Behold, I was shapen in iniquity; and in sin did my mother conceive me" (Ps. 51:5).

Paul's inspired observation in his letter to the Romans reads like a summary of all the newspaper and radio and TV news reports ever given.

Their throat is an open sepulchre; with their tongues

they have used deceit; the poison of asps is under their lips: Whose mouth is full of cursing and bitterness: Their feet are swift to shed blood: Destruction and misery are in their ways: And the way of peace have they not known: There is no fear of God before their eyes. Now we know that what things soever the law saith, it saith to them who are under the law: that every mouth may be stopped, and all the world may become guilty before God (Rom. 3:13–19).

Martin Luther said:

Sin is so deep and horrible a corruption of nature that no reason can comprehend it, but upon the authority of the Scriptures, it must be believed. The original sin is not like all other sins which are committed. The original sin is. It is the root of all sin.[6]

Jesus said:

For out of the heart proceed evil thoughts, murders, adulteries, fornications, thefts, false witness, blasphemies (Matt. 15:19).

Paul explained how the sinful nature of man is manifested in day-to-day living.

Now the works of the flesh are manifest, which are these; Adultery, fornication, uncleanness, lasciviousness, Idolatry, witchcraft, hatred, variance, emulations, wrath, strife, seditions, heresies, envyings, murders, drunkenness, revellings, and such like (Gal. 5:19–21).

Man, then, is capable of doing terrible things. Within him lies the potential of all kinds of crime and violence. The headlines in the daily newspaper and the violence on the evening news are but evidences of what lies within him.

If that is true, why are not all men criminals? Because of certain influences in life that restrain us.

Augustine wrote, "We are capable of every sin we have seen our neighbor commit unless God's grace restrains us."[7]

Christians are a restraining influence on the world. Believers are not all they ought to be, but the world is better because they are here. Jesus referred to His own as the light of the world and the salt of the earth. Those who belong to Christ and are totally committed to Him have a restraining influence on evil, and their presence in the world promotes a higher moral standard among all people whose lives are touched by them. Speaking prophetically, that is why we believe the Antichrist cannot come to power until the church is taken out of the world at our Lord's return.

The Holy Spirit, working through the church, holds back the tide of evil.

> For the mystery of iniquity doth already work: only he who now letteth [hinders] will let [hinder], until he be taken out of the way. And then shall that Wicked be revealed, whom the Lord shall consume with the spirit of his mouth, and shall destroy with the brightness of his coming: Even him, whose coming is after the working of Satan with all power and signs and lying wonders, And with all deceivableness of unrighteousness in them that perish; because they received not the love of the truth, that they might be saved (2 Thess. 2:7–10).

A nation is law-abiding to the extent that its Christian heritage is strong. A revival of a past generation may not continue to fill the churches and produce genuine Christian joy, but some of the influence of that revival may linger: respect for the Bible, an awareness of biblical standards of righteousness, traditions rooted in a more spiritual time, a form of government founded upon biblical principles.

The restraints of home and family have kept many from crime and violence. A husband may restrict his desires because of a deep love for his wife. A wife may overcome temptation to do wrong because of admiration and respect for her husband. Children may refuse to go along in some of their friends' escapades for fear of hurting their parents, whom they love.

Fear of punishment for law-breaking may loom large, deterring the tempted from antisocial behavior.

All these restraints work to hold evil in check and to give society stability. They help keep the lid on crime and immorality. They do not do away with man's potential to sin, but they help to hold him in check, and in this function these restraints are beneficial to all.

Alcohol removes restraints.

The same characteristic of alcohol that allows a drinker to relax at a party eats away at the forces that restrain him from doing wrong. The sections of his brain that are being paralyzed by alcohol, eliminating some of his fears, are the same thinking centers that keep the lion within him caged.

With the entrance of alcohol, the walls fall down. When it penetrates the brain, it lulls the sentries to sleep, allowing man's sinful nature to show itself in behavior.

The defenders of beverage alcohol argue that drinking does not make people commit crimes. They insist that the lawbreaker had crime on his mind or he would not have committed it while under the influence of alcohol.

We could not agree more.

All people have a natural bent to sin. However, God-ordained restraining fences often prohibit the tempted one from carrying out the act. When alcohol removes the restraints, the act is committed. And people suffer. Had alcohol not been used, the crime might never have taken place, and society would have benefited.

Driving down the highway, one might have the urge to swerve from side to side or accelerate to a high speed. Laws act as restraints against such irresponsible driving. But if alcohol is traveling with the driver, toppling the protective walls, he may swerve or accelerate to a dangerous speed and injure or kill himself and others.

One may want to punch another in the nose because of some old grievance, but his better judgment and the fear of prosecution keep him under control. A few drinks can prompt the long-awaited punch, resulting in personal injury, shame, and expense. (Of course, it is also wrong to hold a grievance against someone even if you never take revenge.)

A person under severe financial pressure may be tempted to steal. Restraints may keep him a law-abiding citizen, but enter the beloved enemy and the forces that have held the pressured person steady are nullified. The crime is committed, and legal woes are likely to be added to the already burdened individual.

Along with alcohol's power to remove inhibitions, then, comes its unwanted effect of uncovering and releasing the evil in man. And the results are felt by the entire community. Dr. E. M. Jellinek, formerly of Yale University, wrote:

The death rate, crime rate, or accident rate in a given

community varies according to the average alcohol consumption; and when alcohol [use] decreases so do the death, crime, and accident rates. Relaxation of restrictions on alcohol is followed by a rise in commitments to asylums, hospitalizations, and delinquency.[8]

But what is alcohol's involvement in crime?

The Third Special Report to the U.S. Congress on Alcohol and Health (1978) from the Secretary of Health, Education, and Welfare attempts to answer that question. The report is a bombshell. These new findings indict beverage alcohol as the principle culprit in crime and a number of other serious national problems.

The following is a direct quote from the report.

The relationship of alcohol to criminal behavior varies by type of crime and by the roles of participants in criminal events. Most alcohol-involved violent crime includes both a drinking victim and a drinking offender.

Robbery. Estimated alcohol involvement ranges as high as 72 percent in robbery offenders. Although the vulnerability of skid-row alcoholics to robbery is common knowledge, alcohol used by other robbery victims is relatively unexplored.

Rape. Estimated alcohol involvement ranges as high as 50 percent in sex offenders and 31 percent in rape victims. The most extensive American study on the subject found that in 63 percent of rapes where alcohol was involved at all, both victim and offender had been drinking. Another important finding was that the type and extent of alcohol involvement in rapes was related to the interpersonal relationship of the victim and the offender.

Assaults. Estimates of alcohol involvement in reported assaults vary widely, ranging up to 72 percent of the offenders and 79 percent of the victims.

Homicide. Research based on coroners' reports and detailed case studies suggest that large percentages of offenders and victims had been drinking at the time of the crime. Most studies show that 40 to 60 percent of homicide victims and up to 86 percent of offenders had been drinking when the murder was committed. The presence of alcohol appears to be most likely in homicides where (1) the victim is stabbed, (2) the situation was already violent, and (3) the victim seemed to have precipitated the murder.[9]

This report reveals a dimension to alcohol's destruction not often considered: the drinking victim as a causative factor in crime. Evidently, nearly one third of the reported rapes in America would not have taken place had the victim not been drinking. It appears that nearly eighty percent of the reported assaults would have been avoided had the victims not been using alcohol. Most important, the report seems to indicate that nearly half of the murders would not have taken place had the drinking victims not precipitated the crime.

The report goes on to say that as many as eighty-three percent of offenders in prisons or jails reported alcohol involvement in their crimes. (While drinking often has been thought to be associated more with crimes against persons than crimes against property, prison data indicates that drunkenness is just as frequent in both.) The government publication "Prisoners in State and Federal Institutions," published December 31, 1974, found that an estimated forty-three percent of all inmates under the jurisdiction of state correctional authorities reported they had been drinking at the time of their offense. Evidently, in four years the figure nearly has doubled, reflecting a greater increase in drinking among those who commit crimes than in the general population.

What awful destruction the enemy alcohol brings to our land! In the area of reported crime alone, if the use of alcohol had been eliminated last year, we might have had only twenty-eight percent of the thefts, about half of the rapes, only twenty-eight percent of the assaults, and but fourteen percent of the murders.

How does this translate into human loss?

Who can measure the amount of grief felt by victims and their families?

And what will we learn from this information?

If past experience is a guide, criminologists and politicians will explore every avenue of combating crime except the one that could make a difference.

New studies will be made. More prisons will be constructed. More police officers will be hired. More judges will be installed. More citizens' groups will be formed.

And the enemy will be untouched.

Why? Because too many of the policymakers share the love affair with alcohol and are unwilling to touch their beloved.

To try to fight crime without facing the alcohol issue is to work blindfolded. Even if we had the most ideal system of corrections, we could not stop crime's onslaught because of alcohol's devastating effects. If we had the most stringent sentencing coupled with perfect police work and flawless penology, the best we could do would be to deal with a fraction of the total crime problem, if the issue of alcohol remained unchecked. Most crime would keep surging on year after year.

The December 26, 1976, issue of the Chattanooga

News Free Press carried an article entitled, "Alcoholism: Big Problem for Alaska." Associated Press writer Tad Bartimus sets a Saturday night scene in Nome, Alaska.

Even for a Saturday, it had been a rough night. The jail was busy. The hospital was busy.

About 2 A.M., the police brought in one fight's drunken loser who was so banged up it took the nurses a few minutes to discover his nose was missing.

Half an hour later, a police officer showed up with a shredded appendage, and the weary surgeon went to work.

Alcoholism in all its accoutrements—crime, suicide, just plain trouble—are the bane of Alaska, and especially northwest Alaska where the alcoholism rate is one of the highest in the world and still growing.

Alaska leads the nation in per capita alcoholic intake—3.86 gallons a year for everyone over 15. One of every 10 people in a state with 360,000 population is an alcoholic or a problem drinker. The problem is especially bad among Eskimos (nearly 20 percent of the total population) in small bush towns like Nome on the Bering sea coast, only 120 miles from Siberia.

Perhaps it is that day and night lose meaning in the land of the midnight sun.

Last year 1,277 drunk and disorderly arrests were logged in this old gold mining town. It only has 2,500 residents.

Police Chief Cecil Johnson says 99 percent of their emergency cases involve alcohol. District Court Judge Ethan Windhal says: "If it weren't for demon rum, I wouldn't have a job."[10]

The facts seem to indicate that Judge Windhal would not be the only public official in the criminal justice system out of work if alcohol use dropped significantly, a development that is long overdue.

But could our nation survive the financial impact of a meaningful drop in the liquor business?

What would happen to our economy?

How would the government continue to operate without liquor revenues?

Wouldn't taxes have to be increased?

Doesn't alcohol pay its way?

6

THE REAL COST OF ALCOHOL

If you are hospitalized in Minnesota, there's a good chance your temporary home will have been partially paid for by tax dollars provided through the sale of beverage alcohol. The sewers outside the building and the water mains bringing that clear Minnesota water to your room are likely to have been made possible by the same benefactor.

Travelers drive through Texas on roads constructed in part with money brought in through the sale of booze.

In Utah, alcohol revenue helps provide school lunch programs.

The Children's Colony in Arizona is financed by taxes from the sale of alcoholic beverages, as are many of the playgrounds in Nashville, Tennessee.

Medical and biological research at Washington State University is made possible by money produced from liquor taxes. Interestingly, research there has recently pointed out the seriousness of the fetal alcohol syndrome.

Revenue from the sale of alcohol in the United States totals approximately ten billion dollars annually. The federal government's share of this amount is somewhat more than that of the states'. For example, in 1975 total revenue was $9.6 billion, with $5.4 billion going to the federal government and $4.2 billion going to the states.[1]

Tax revenue is not the only way that beverage alcohol benefits the nation. One out of every forty-nine working Americans is employed by the liquor industry, and more than ten billion dollars is paid yearly to these workers.[2]

Taxation of alcoholic beverages for the support of public projects is an old practice in America, dating back to colonial times. It began in Massachusetts, and by 1750 nearly all the colonies had joined the parade. The money received was used then primarily for maintaining the state militia. In 1727, Connecticut designated its revenue from rum sales for education, with a good amount of it going to support Yale University in Maryland. Some colonies also used alcohol taxes for the operation of courts and jails.

The Civil War brought national taxation of beverage alcohol under the Internal Revenue Act, passed as an emergency measure to raise money to carry on the war. This law placed a federal tax by the gallon on the manufacture of liquor, beer, and ale; it also required a federal fee of every retail liquor establishment in the Union. The law was supposed to be repealed at the end of the war, but like most taxes, it endured long after the last shot was fired in that tragic conflict, remaining in effect until 1915.

Until the enactment of the federal income tax, liquor revenue provided nearly two thirds of all the federal government's income. It is unlikely that prohibition would have become law without the introduction of the federal income tax. During prohibition, of course, there was no tax income from liquor sales, and after its repeal liquor revenue never became a vital part of federal income.

Does this mean that ten billion dollars of annual liquor revenue is not important? Important, perhaps, but not crucial. The total amount of revenue from alcohol taxation makes up only 1.9 percent of federal income and an equal amount for the states.

In view of the total tax picture, alcohol pays a very small amount. And with taxes on alcohol remaining constant and liquor prices relatively stable, the percentage of the national tax burden born by the producers and sellers of beverage alcohol lessens year after year because of inflation. A report in the magazine *The Bottom Line* reveals the following:

> Since 1970, the overall CPI (Consumer Price Index) has risen by an average of nearly 10 points per year. During the same period, prices for distilled spirits have gone up by only 2 points annually.
>
> One reason for the relative stability of alcoholic beverage prices over the past decade has been a freeze on federal tax rates. The last increase at the federal level for beer, wine, or distilled spirits was in 1951.
>
> The net result has been for liquor to be cheaper in terms of the consumer's disposable income, cheaper than food, housing, or most other items figured in the Consumer Price Index.[3]

If this seems irrelevant or a bright spot in the economy, think again. The more reasonably priced alcohol is in relation to other products, the more available it is to the general public. Greater availability translates to greater use, and in turn to greater destruction. Clearly, it is time to change the tax structure on all alcoholic beverages, bringing these products into line with other items and causing this industry to bear its share of the budget burden it creates.

Taxes collected on booze do not really enrich the common treasury. This is a case where it is wise to look a gift horse—alcohol—in the mouth. Social problems growing out of alcohol use are expensive. And revenues received through taxation of alcoholic beverages do not cover the cost of the trouble created by this enemy.

Consider these revealing reports:

In 1943 the General Court of Massachusetts reported that for every dollar of beer and liquor tax received, the Commonwealth of Massachusetts spent $3.50 for known and measurable costs. In 1957 a survey by the Utah State Board on Alcoholism disclosed that for every dollar of beer and liquor tax received, the state was spending $1.37 just to care for alcoholics. In 1971 a study by economist W. Slater Hollis indicated that for every dollar of liquor revenue that Tennessee received from Shelby County and Memphis, it cost the state $2.28 in alcohol-related expenses. . . .

In a new study prepared under a grant from the NIAAA (National Institute on Alcohol Abuse and Alcoholism), researchers Berry, Boland, Smart, and Kovak estimated the economic impact of alcohol problems in the U.S. for 1975 at $42.75 billion.[4]

The breakdown of this $42.75 billion expense is as follows:[5]

Lost Production	$19,640,000,000
Health and Medical	12,740,000,000
Motor Vehicle Accidents	5,140,000,000
Violent Crime	2,860,000,000
Social Responses	1,940,000,000
Fire Losses	430,000,000
TOTAL	$42,750,000,000

Remember that these alcohol-related expenses were for 1975. In that same year, revenues received from liquor sales were $9.6 billion. In 1975, then, for each $1.00 received in taxes on alcoholic beverages Americans paid $4.41 to care for alcohol problems. Any way you figure it, the booze business is bad business for all Americans, except those directly profiting from that industry.

But the tax tale is not the whole story.

Nearly $20 billion in lost production translates into higher prices for everything. This industrial hangover accelerates the spiral of inflation and eats away at the economic stability of our nation. Alcohol use by employees results in higher prices and poorer merchandise. Every American who writes a check for any product is paying part of the booze bill.

New automobiles cost more because of alcohol.

Television sets cost more because of alcohol.

The clothes you wear cost more because of alcohol.

All manufactured goods are more expensive because of the cost of alcoholism in industry.

Health and medical costs are out of sight. One of the reasons is the rampant use of the drug alcohol. One dollar in every five spent for hospital care in the United States is for an alcohol-related illness or problem.[6]

The health field is sick economically. Soaring costs have reached a critical state. Socialized medicine is impending. If that happens, freedom in the medical area may soon be lost, and alcohol will have been one of the thieves of liberty.

Motor vehicle accident costs affect all people. More than five billion dollars in losses must be spread over the entire insurance industry, and each of us is forced to pay a part. You may be fortunate enough to avoid the drinking driver swerving down the highway, but alcohol will get you in the pocketbook when it comes time to pay your automobile insurance premium. No American driver escapes.

Although you may have an excellent driving record and be favored with lower rates because you do not drink, the possibility of encountering a drinking driver must be figured into your insurance costs. You lose when others drink—even though you may not be involved in a traffic accident.

Crime costs. And much of the crime in America is alcohol related.

Packed prisons are common in nearly every state. The cost of imprisoning a person for one year is estimated at approximately twenty thousand dollars. *U.S. News and World Report* has projected prison building needs throughout the nation at $4.7 billion by 1985. And that staggering amount represents only the construction costs! Imagine the expense to taxpayers in meeting payrolls for additional prison personnel. The most conservative estimates recognize one fifth of all criminal justice costs to be alcohol related. More realistic figures may double or even triple that amount.

Another expense connected with the prosecution and imprisonment of criminals is the amount of welfare required to care for their families. Many millions of dollars are provided by law-abiding citizens to meet the needs of convicts' dependents, and alcohol has been the partner in crime of thousands who are now behind bars, unable to care for their families.

Social responses include all types of programs designed to aid those struggling with alcohol. It should be noted that about twenty percent of all money received in tax revenue on alcoholic beverages is spent in social responses to the problem. And nearly all authorities feel that the present effort is feeble and far too small. Future expenditures in this area will rise dramatically as government agencies try to hold down alcohol abuse. And since booze is inadequately taxed, drinking and non-drinking taxpayers alike will have to pay the bill for these programs.

Actually, there is no program for alcoholics as successful as the simple application of the gospel of Christ. Still, while missions to alcoholics struggle financially, Americans continue their booze binge, spending nearly three times as much for alcoholic beverages as for all religious and charitable purposes.

There is another dimension to the cost of alcohol use that cannot be measured: the cost of human suffering. Consider the grief of those who lose loved ones in automobile or industrial accidents; the hurt of those who experience the heartbreak of divorce; the physical agony of those who endure disease and pain as a result of alcohol's impact upon their bodies, and the mental anguish of those who are pushed to the brink by alcohol problems of their own or of those close to them.

The list goes on.

Alcohol costs.

Writing in *The Church Herald,* Dr. Willard Brewing said:

The liquor business is the only business I know that keeps a one-sided ledger—that deals with receipts and ignores legitimate expenditures. It takes the profits and leaves the expenses to someone else. Within one year in Vancouver I buried three young people who had been killed by alcohol on the highways. That is not my assumption; that is the verdict of the jury. At not one of those funerals was there a representative of the business that killed them. They did not pay the funeral expenses or look after the orphans. If it had been a railway accident, the fault of the road, they would have been there. Had it happened in an industrial plant, they would have been there. But this accursed business grasps its privileges but takes no stock in its obligations. Some foolish people seem to think it pays. If it met its legitimate bills, it would go bankrupt in a year.[7]

And so it would. The liquor industry cannot afford to pay for the havoc its product creates. Nor can we.

But the public cost for alcohol's destruction does not compare to the loss its impact brings to individuals and families. Here the trade-off of booze for food, clothing, and education is a continuing tearful drama of deprivation.

At one time in our history, the American people became so aroused over their costly experience with beverage alcohol that they were moved to action.

It is time for that righteous ire to rise again.

7

GREAT BOOZE FIGHTERS OF THE PAST

America's problem of alcohol addiction is not a phenomenon of the twentieth century. Those who long to look back to a "better day" will find little in the early history of the nation to encourage them. Beverage alcohol came with the first immigrants.

When John Winthrop, the first governor of the colony of Massachusetts, arrived aboard the *Arabella*, the ship carried forty-two tons of beer.

Henry Hudson piloted his famous ship, the Half-Moon, into the mouth of a river that he hoped would be the Northwest Passage. He stopped at a flat, wooded island and debarked from the ship to try to make friends with some Indians who were fishing there. Upon meeting them and communicating by sign language, he offered them drinks from the cask of gin he had brought ashore. The Indians were so taken with this new beverage that they later named the island *Manahachta-nienk*, which means "the place where we all got drunk." The name was later shortened to Manhattan.

Although alcohol use was common in early America, the colonists were harsh on those given to heavy drinking. The first offense would probably result in some time in the stocks or on the whipping stool. On the second offense, a man might be sentenced to wear a "D" around his neck for one year so everyone would know he was a drunkard.

Dr. Benjamin Rush, a signer of the Declaration of Independence and the surgeon general of the Revolutionary army, was one of the first important Americans to speak out on the evils of alcohol use. Rush had read a pamphlet by Anthony Benezet, a Quaker intellectual, entitled, "The Mighty Destroyer Displayed" that exposed the detrimental effects of beverage alcohol. Benezet's pamphlet did not have wide distribution, but it did impress Rush, who in turn influenced the thinking of Americans about alcohol for the next hundred years.

When Rush assumed his duties as surgeon general of the Revolutionary army, he discovered that drinking was doing more harm to the American cause than was the British army. Alcohol also was used freely in the army as a medicine for a number of illnesses. Rush objected to this practice, becoming the first American in a high government position to protest the use of alcohol for medicinal purposes. He insisted that liquor was detrimental to the army and to society, associating it with crime, disease, suicide, and death.

After the Revolutionary War, Rush pursued his research on beverage alcohol and was so moved by his findings that he called for ministers of all the churches to aid him in the battle against its use. In his words, the call was to "save our fellow men from being destroyed by the great destroyer of their lives and souls." He also published an article entitled, "Directions for Preserving the Health of Soldiers" that warned against liquor use.

In 1784, Rush wrote a pamphlet entitled, "Inquiry Into the Effects of Spiritous Liquors on the Human Body and Mind." The pamphlet was a bombshell. Issues of the publication were exhausted as fast as they came off the press, and demand continued for thirty years. Long after Rush died, his pamphlet went on fighting the battle against alcohol use. Preachers and temperance speakers quoted freely from it for decades. So powerful was Rush's argument against liquor that some distillers voluntarily went out of business after reading his pamphlet.

While opposition to alcohol use in America began to build as a result of Rush's research, the enemy continued to strengthen its hold on the young nation. In 1810, drinking was rampant. Per-capita intake of alcohol was even more than it is today. Booze was served at gatherings of almost every kind. In 1842, Abraham Lincoln recalled his childhood as follows:

> ... we found intoxicating liquor ... used by everybody, repudiated by nobody. It commonly entered into the first drought of the infant, and the last drought of the dying man. From the side board of the parson, down to the ragged pocket of the homeless loafer, it was constantly found. ... To have a rolling or raising, a husking or hoe-down, anywhere without it, was positively insufferable.[1]

Lincoln added that during those years the devastator (liquor) had come forth across the land, "Like the Egyptian angel of death, commissioned to slay if not the first, the fairest born of every family."

Historian Henry Adams said of that time that nearly every American family, regardless of their social position, had some member who was the victim of intemperance.

One of the reasons for alcohol's early hold on the young nation was the lucrative trade built up involving rum, molasses, and slaves. Writer Donald Barr Chidsey explains:

> By the early nineteenth century there were forty distilleries in Boston, twenty-one in Hartford, and eight in Newport, all making rum. Some of the rum was sold in the colonies, some was transported to Europe, and a generous quantity was drunk at home. The rest of the rum was shipped to the Guinea Coast to pay for more slaves, who were then taken to the West Indies and traded for molasses, which was taken to New England and made into rum, which was sent to the Guinea Coast to purchase slaves. . . .So it went for many years. It was a vicious but lucrative triangle. Some of the stuffiest New England families owe their current affluence to this trade. . . .
>
> It was largely because of this cunning commercial arrangement that "rum" came to be a generic word in the United States for all hard liquor, as in phrases like, "demon rum," "rum pot," "rum row." It is not so used anywhere else in the world.[2]

In addition, settlers inland found it more profitable to make their grain into whiskey and transport that for sale to the more populous areas than to transport the grain itself. The newborn nation then experienced double trouble: rum on the coast, whiskey inland, and beer and hard cider plentiful almost everywhere.

The old-time saloon became the local watering hole for those with a thirst for strong drink. The atmosphere of those centers of immorality, lewdness, and drunkenness is almost indescribable. There, unprincipled men met to hatch plots that would give them financial advantage over others. Crooked politicians found the saloon a natural place to further their aims. Prostitution was often carried on in connection with a nearby hotel, where the saloon management would arrange for whole floors of rooms to be available for customers. Gambling, violence, and drunkenness were everyday occurrences in the saloons that were multiplying across the nation at an alarming pace.

In his book *Deliver Us From Evil*, Norman H. Clark says the saloon was a commerce not only in beverages, but in flesh and corruption. He writes:

> The saloon business could degenerate into a competitive frenzy which might infect even a town of only a few hundred citizens with dozens of dingy saloons. The competition was to lure customers into drinking as much as they could possibly hold. After the famous—and notoriously salty—"free" lunch, the thrust toward competitive advantage often lay through prostitution, gambling, and, later, narcotics. In a broader and more insidious fashion the liquor traffic presented a protective facade for organized criminal activity, which was often organically linked to the saloon.
>
> And when they were pressed to defend themselves, the liquor interests for decades herded blocks of semiliterate voters, tyrannized political party conventions, bought elected officials, then diligently frustrated the forces of law, order, and reform. As the traffic became national and industrial, the big breweries which owned about 70 percent of the saloons, needed the local business licenses that only political influence could provide, and they had the kind of fluid cash incomes that corrupt politicians could not resist. At

every level of municipal or state responsibility, the liquor interests polluted politics and fought viciously against efforts to restrict their adjunctive relationships to prostitution and gambling. They repeatedly financed campaigns in opposition to women's suffrage, and they did their best—they performed at their worst—in perverting Democratic practice in the many state and local referenda on the question of licensing saloons.[3]

Long ago the Hebrew prophet Isaiah had written, "When the enemy shall come in like a flood, the Spirit of the LORD shall lift up a standard against him" (Is. 59:19). The flood of alcohol that threatened to destroy the nation caused a reaction among concerned Americans that significantly reduced the use of alcohol, and finally ended the sale of legal liquor in the land, for a time.

Excess and iniquity resulted in indignation.

Individuals and groups rose up to stand against the liquor traffic. For a century, the United States was a battleground between those who sold beverage alcohol and those who were part of a growing grassroots movement that opposed its use.

Dr. Billy J. Clark practiced medicine in a small town in upstate New York. He might have lived and died in obscurity had it not been for his reading of Rush's pamphlet on the effects of alcohol. His daily medical work brought him in contact with lumber workers and farmers who were heavy drinkers, and he witnessed firsthand the devastating effects of alcohol on his patients.

Clark finally became so concerned about the damage being done in his community by liquor drinking that he enlisted the help of his pastor and called a meeting of forty influential men in the town to find a solution to the problem. They met on April 30, 1808, and organized the Union Temperance Society of Moreau and Northumberland. This was the first temperance society in America. The members each pledged to abstain from any distilled spirits for one year, hoping that others would follow their example.

The community was shaken.

Dire predictions were made about the loss of labor that would come without the free flow of spirits. Prophets of doom insisted that nearly all joint work efforts such as husking corn, building houses, and making quilts would cease. Opponents of the plan expected the general health of the area to decline, since alcohol was considered medicine for nearly every ailment of the day.

At the end of the year, however, the Union Temperance Society of Moreau and Northumberland was going strong. Members found that avoiding alcohol had been beneficial, increasing their efficiency during the year and giving them a better quality of life. A public meeting was held in which members shared their experiences. Newspapers picked up the story and other communities began to form their own temperance societies. Some groups were small; others gathered members throughout entire states. The move was on.

Rev. Lyman Beecher was the pastor of a church in Litchfield, Connecticut. Upon reading Rush's pamphlet on alcohol, he said, "It fermented in my mind." Beecher was disgusted by the use of liquor among preachers and the general public, and he decided to do something about the situation.

In 1825, speaking from his pulpit in Litchfield, he began a series of sermons on the evils of drinking. After the sermons had been delivered, he put them into print under the title, "Six Sermons On the Nature, Occasions, Signs, Evils, and Remedy of Intemperance."

It was as if the nation had been thirsting for Beecher's information. The publication of his sermons ran through five printings in five months, and enjoyed wide circulation for ten years.

In 1826, he became the pastor of Hanover Street Church in Boston. In this larger church he was even more effective in helping to form a national movement opposing the use and sale of liquor.

Beecher's daughter, Harriet, was also an able writer on the subjects of temperance and abolition. She became well known for her novel *Uncle Tom's Cabin*. The temperance and anti-slavery movements found a great deal in common and worked hand in hand prior to the Civil War.

One of the most effective anti-liquor organizations started in a tavern. Six men composed the entire membership of a small drinking club that met weekly to imbibe and enjoy one another's friendship. These men were serious drinkers, and alcohol was always the center of their weekly gathering. They were James McCurley, a coachmaker; William K. Mitchell, a tailor; John F. Hoss, a carpenter; Archibald Campbell, a silversmith; and David Anderson and Gordon Stears, blacksmiths.

On April 2, 1840, the six drinkers were meeting as usual at Chase's Tavern in Baltimore. On that same night Matthew Hale Smith was speaking at a nearby church on the subject of temperance. Not wanting to interrupt the drinking of the entire group, they sent two of their number to hear the temperance speaker.

When the two delegates returned, they had been so moved by Smith's message that they enthusiastically shared his views with the rest of the club. What had begun as a joke or a drinker's whim turned out to be serious business. There in Chase's Tavern, after a drinking bout, these six men agreed to organize a total abstinence movement. They called their new group "The Washingtonians," named after the first president of the United States. At their next meeting, one week later, they adopted the following pledge:

> Whereas, the use of alcoholic liquors as a beverage is productive of pauperism, degradation, and crime and believing it is our duty to discourage that which produces more evil than good, we therefore pledge ourselves to abstain from the use of intoxicating liquors as a beverage.

The Washingtonians used public confession to spread their message. Their gatherings were known as "experience meetings." The aim at each meeting was to bring heavy drinkers who could be told about the benefits of abstaining from alcohol. Washingtonians had no desire to form a political action group or even to make booze illegal. In fact, they felt that abstinence should be voluntary. Nevertheless, they were serious about their efforts to free men from the chains of drink. By 1847, their membership had grown from 6 to 600,000.

Neal Dow, a prohibition pioneer, was born in 1804 in Maine. In his youth he was exposed to drunkenness in a neighbor's home and in the general area in which he lived. He became a successful and wealthy businessman while still very young, and after observing the effects of alcohol upon its victims he decided to use his energies and abilities to fight the liquor traffic. He declared: ". . . the traffic in intoxicating drinks tends more to degradation and impoverishment of the people than all other causes of evil combined."[4]

In the early 1840s Dow had an experience that launched him on a successful crusade to bring prohibition to Maine. With his help, a poor relative who had a drinking problem found a job. Dow then went to the local saloon and asked the owner to refuse to sell drinks to this relative for the sake of the man's family. The owner growled that he was running a legal business and that he would serve whom he pleased. Dow was furious at the barkeeper's lack of concern and determined: "With God's help I will change all this."[5]

He was as good as his word.

Through untiring work and dedication, Neal Dow brought the first prohibition law in America to his own state. In 1851, Maine went dry, largely through his efforts. By 1855, Dow had crusaded widely and had brought prohibition to Minnesota, Rhode Island, Massachusetts, Vermont, Michigan, Connecticut, Indiana, Delaware, Iowa, Nebraska, New York, and New Hampshire—thirteen states in all. Some of these newly enacted prohibition laws were declared unconstitutional by the courts, and considerable progress was lost for the cause during the Civil War. Later in the century the battle with alcohol would be fought again. Neal Dow would not be forgotten as one of the early champions of prohibition.

Abraham Lincoln raised a strong voice against alcohol. In a temperance address in 1842, he capsulized the issue in a manner that needs to be repeated again and again in our day. Lincoln insisted that the social and personal disasters brought by liquor came not from "the abuse of a very good thing," but from "the use of a very bad thing."

On the day of his death Lincoln said:

> With the help of the people, we have cleaned up a colossal job. Slavery is abolished. After reconstruction,

the next great question will be the abolition of the liquor traffic. My head and heart and my hand and my purse will go into that work. Less than a quarter of a century ago I predicted that the time would come when there would be neither a slave nor a drunkard in the land. I have lived to see, thank God, one of these prophecies fulfilled. I hope to see the other realized.[6]

Disrupted by the Civil War, the temperance movement lost ground for a time. Since many booze fighters were also slavery fighters, the times demanded that their energies be given to the issue at hand. By 1869, however, the National Prohibition Party had been formed, and shortly thereafter women across the nation became more active than ever before in the battle against beverage alcohol.

In 1873, Dr. Dioclesian Lewis, a prolific author of health books and a public speaker, gave a lecture on temperance in Hillsboro, Ohio. This city of three thousand had thirteen saloons and a drugstore that sold liquor. Lewis related an experience from his youth, when the women in his home town marched upon the saloons singing hymns and praying, and drove the liquor sellers out of business. At the end of his lecture, he asked how many of the women present would be willing to join in such a venture. A good number of the women responded, launching what would be known as the Women's Crusade.

The following day a sizeable group of women set siege to the saloons in Hillsboro—singing, praying, weeping, and urging the saloon owners to destroy their booze and go out of business. Although they were unwelcome, they knelt on the sawdust floors of the saloons and poured out their supplications to God. They met for prayer meetings in the morning, marched on the saloons during the day, and held mass meetings in the evening. Within two months, nine of the thirteen saloons had closed. Some had gone bankrupt. Others had simply surrendered to the attack.

The crusade caught fire, spreading to hundreds of towns and cities in Ohio and surrounding states.

The women involved in the crusade underwent severe persecution. Saloonkeepers and their customers were determined to let these invaders know they were not welcome. Saloon doors were locked against them. Men tried to drown out their singing and praying with mockery and obscenities. Some of the women were beaten and subjected to indignities. Often they were arrested and jailed. They were threatened, and their leaders were hanged in effigy.

Brass bands were hired to gather around the women and play as loudly as possible so that none could hear their singing and praying. Booze was poured over them. But the characters of the saloonkeepers were exposed to public opinion, as was the generally consistent Christian attitude of the marching women. Thousands of saloons went out of business as a result of the Women's Crusade.

The Women's Christian Temperance Union (WCTU) began in 1874. It probably was born as a national organization as a result of the Women's Crusade. Under the direction of Frances Willard, the WCTU became a powerful force in moving the nation toward prohibition.

The Anti-Saloon League was founded by Dr. Howard Russell, an Ohio pastor, and was organized on a national basis in 1896. Russell and his associates called their organization "The Church in Action Against the Saloon." More than any other, it was the vehicle used to successfully win the battle for national prohibition. The League was able to unite most of the various temperance groups across the nation, and was so successful that at one time they were able to offer fifty thousand speakers to be used for the cause anywhere in the land. The Anti-Saloon League was highly organized and ultimately exerted immense political power. Most of its financial support came from churches, and a considerable amount was spent on printing millions of pieces of literature setting forth the League's case against alcohol.

At the turn of the century there were three dry states in the nation. For more than a century the battle had raged. There had been gains and losses, times of discouragement and disorganization, and other times of superb planning and advancement.

Although there were fewer dry states than at some times in the past, the anti-liquor forces saw their brightest hour just ahead. Public sentiment against the evils of alcohol was high, and the Anti-Saloon League had been able to pull the dry forces together and gain political advantage on a number of fronts. William Jennings Bryan, three times a candidate for president of the United States, and secretary of state under President Wilson, proved an eloquent spokesman with a broad following throughout the land.

All the movement lacked was a fiery speaker who could move people at the grass-roots level. Such a spokesman appeared, walking off a baseball diamond. After drinking with other ball players in a Chicago saloon, Billy Sunday was converted to Christ in the famous Pacific Garden Mission.

Billy Sunday made war with the devil. And he looked upon booze as the devil in solution, the saloon as the center of wickedness. Hear him:

The saloon is a liar. It promises good cheer and sends sorrow. It promises health and causes disease. It promises prosperity and sends adversity. It promises happiness and sends misery. Yes, it sends the husband home with a lie on his lips to his wife; and the boy home with a lie on his lips to his mother; and it causes the employee to lie to his employer. It degrades. It is God's worst enemy and the Devil's best friend. . . .It spares neither youth nor old age. It is waiting with a dirty blanket for the baby to crawl into this world. It lies in wait for the unborn.[7]

Billy minced no words. He called the saloon the sum of all villainies, the parent of crimes, and the mother of sins. He declared, "I am the sworn, eternal, uncompromising enemy of the liquor traffic. I ask no quarter, and I give none."[8]

Sunday held great crusades in cities across the land. Huge tabernacles were built for his meetings. He met the people where they were, spoke their language, and gave them a message that went well beyond the liquor problem. While speaking plainly against booze, his aim was to convert people to Christ, expecting this miracle to change their views on such issues as the question of alcohol. His approach was spiritual, biblical, and effective.

Billy Sunday was the man of the hour. He had been brought to the kingdom for such a time as this.

When prohibition became the law of the land, Sunday staged a funeral service in Norfolk, Virginia, for John Barleycorn. The long struggle seemed to be over, and Billy made the most of it. At one minute past midnight on January 17, 1920, the "body" of John Barleycorn (whose name was synonymous with booze) was carried in a twenty-foot horse-drawn casket to

Billy's tabernacle. He began his long-awaited funeral sermon with these words, perhaps summing up the hopes and visions of all the booze fighters of the past century:

> Good bye, John. The reign of tears is over ... the slums will soon be a memory. We will turn our prisons into factories and our jails into storehouses and corncribs. Men will walk upright now, women will smile, and the children will laugh. Hell will be forever for rent.[9]

To what degree did national prohibition fulfill the booze fighters' dreams?

What really happened during prohibition?

8

PROHIBITION: WHAT REALLY HAPPENED?

Prohibition didn't come to America overnight.

More than a century of struggle went into the fight to make the nation dry. Finally, an indignant populace, tired of the evils of the liquor trade and the crime-producing neighborhood saloon, decided to throw off the chains of legal alcohol and accept national prohibition.

Despite the long period of travail in bringing prohibition to America and the rich source of history this period provides, most Americans today know little about the prohibition era. They are likely to think of it as an experiment that lasted only a few years, during which drinking was heavier than ever and gangsterism was the order of the day.

Opponents of prohibition have long claimed that the Eighteenth Amendment was voted in while a million American soldiers were in Europe during the First World War.

The charge is ridiculous. After the turn of the century a great tide of prohibition rolled in and covered the land. By 1917, thirty-three states had adopted state-wide prohibition. National prohibition began on January 17, 1920, and it was adopted with such enthusiasm and by such a great margin that all of the votes of the men overseas could not have changed the outcome.

In September, 1923, Felix Frankfurter, a justice of the U.S. Supreme Court, set the record straight on this charge in the annals of the Academy of Political and Social Science:

> It is sheer caricature to convey the impression that the 18th Amendment came like a thief in the night. Prohibition was the culmination of fifty years of continuous effort; nor did the movement lack alert, persistent, and powerful opposition. If the process by which this amendment came into the Constitution is open to question, one hardly dare contemplate the moral justification of some of the other Amendments to the Constitution itself.[1]

Liquor propagandists have done their work well. Ask most people what happened during prohibition, and they will tell you the following:

    1) It was a time of heavy drinking, heavier than when alcohol was legal.

    2) It was a time when crime was rampant and gangsters ruled the cities, a time of political corruption.

    3) It was a time of economic depression.

But what really happened during this time?

Prohibition had been adopted to reduce the public's use of alcohol, and in this it was successful. The consumption of alcohol during prohibition, especially at its beginning, dropped drastically. Among working people, alcohol use was cut approximately in half. Drunkenness was almost nonexistent.

While there are no records to show the amount of illegal alcohol that was consumed during prohibition, researchers have been able to determine the general amount of drinking in that period by studying illness and social problems associated with beverage alcohol.

For example, the percentage of deaths in the United States caused by cirrhosis of the liver for the years 1900 through 1917 was 13.11 percent. From 1920 through 1932 (all prohibition years), this rate dropped to 7.27 percent. After prohibition ended, this figure began to rise, and by 1965 it had reached 12.5 percent.[2] In the 1970s, incidence of this dreaded disease exceeded the pre-prohibition rate.[3]

Deaths from Bright's disease, pneumonia, and tuberculosis dropped dramatically during prohibition, and admissions to mental hospitals for alcoholic psychoses fell to their lowest point in history.

Writing in the *Journal of Social History*, J. C. Burnham explained:

> Undoubtedly, the most convincing evidence of the success of prohibition is to be found in the mental hospital admission rates. There is no question of a sudden change in physicians' diagnoses, and the people who had to deal with alcohol-related mental diseases were obviously impressed by what they saw. After reviewing recent hospital admission rates for alcoholic psychoses, James V. May, one of the most imminent American psychiatrists, wrote in 1922: "With the advent of prohibition the alcoholic psychoses as far as this country is concerned have become a matter of little more than historical interest. The admission rate in New York state hospitals for 1920 was only 1.9 percent [as compared with 10 percent in 1909–1912]." For many years articles on alcoholism literally disappeared from American medical literature.[4]

Connecticut was one of the states that voted against the Eighteenth Amendment. Nevertheless, the people of Connecticut benefited from prohibition when it became a national law. In his book *The Law of the Land and Our Moral Frontier*, Prof. Henry W. Farnam of Yale University lists the following effects of prohibition in Connecticut:

1) Prisoners in Connecticut jails charged with drunkenness fell from 7,314 in 1917 to 943 in 1920.

2) Arrests for assault and breach of peace declined to less than one third.

3) Jail commitments for vagrancy practically vanished.

4) Commitments for alcohol insanity in 1920 were less than one third the number of 1917.

5) Death from alcoholism and cirrhosis in 1920 was less than one half the 1917 rate.

6) Accidental deaths in 1917 were 10.7 per 10,000; in 1920 it was 7.3.

7) Automobile death rate fell by 40 percent.

8) Death rate from tuberculosis dropped from 15.3 per 10,000 to 9.6 in 1921.

9) Death from pneumonia, to which alcoholics are particularly liable, fell by over 50 percent.[5]

Alcohol's unmistakable trail of destruction was nearly absent during prohibition. Research reveals that legends about increased drinking during prohibition have no basis in fact. Norman H. Clark concludes:

There are today few reasons to believe that these legends, even those so recently embellished, are more than an easy and sentimental hyperbole, crafted by men whose assumptions about a democratic society had been deeply offended. . . . To suppose, further, that the Volsted Act (Prohibition) caused Americans to drink more rather than less is to defy an impressive body of statistics as well as common sense. The common sense is that a substantial number of people wanted to stop both their own and other people's drinking, and that the saloons where most people had done their drinking were closed. There is no reason to suppose that the speakeasy, given its illicit connotations, more lurid even than those of the saloon, ever, in any quantifiable way, replaced the saloon. In fact, there is every reason to suppose that most Americans outside the larger cities never knew a bootlegger, never saw a speakeasy, and would not have known where to look for one.[6]

Why, then, did the rumors arise, and why do the legends continue?

When liquor became illegal and expensive, only the affluent could afford it. Drinking then became a status symbol. And while the well-to-do had used little alcohol previous to prohibition, patronizing bootleggers became "the thing to do." So while the drinking habits of the masses were reduced considerably during prohibition, many affluent people began to drink more heavily. Since these people were far more visible to journalists and other people-watchers, the legend was born and has been perpetuated.[7]

Perhaps the most convincing proof of prohibition's success is to be found in a comparison of the use of alcohol before and after prohibition. Government reports show that in 1914 the per-capita use of alcoholic beverages was 22.80 gallons. In 1934, the first year after repeal, the amount was 8.96 gallons. In a sense, the nation had been weaned away from its drinking habits during the nearly fourteen years of prohibition. It took many years of promotion by the liquor industry to bring drinking in America up to pre-prohibition levels. Beer production did not reach that point until

1943, nearly ten years after the end of prohibition.[8] Today, per-capita use surpasses even that of the pre-prohibition era, standing at nearly ten gallons above the level that raised the ire of Americans to the point of bringing about national prohibition.

Did prohibition create a crime wave in the land? Absolutely not.

In 1922, Charles W. Elliott, president of Harvard University and a lifelong proponent of prohibition, wrote the following to the Massachusetts legislature:

> Evidence has accumulated on every hand that prohibition has promoted public health, public happiness, and industrial efficiency. This evidence comes from manufacturers, physicians, nurses of all sorts (school, factory, hospital), and from social workers of many races and religions, laboring daily in a great variety of fields. Testimony also demonstrates beyond a doubt that prohibition is actually sapping the terrible forces of disease, poverty, crime, and vice; in spite of imperfect enforcement. It has eliminated the chief causes of crime, poverty, and misery among our people.[9]

Did not the well-known gangs of the prohibition era capitalize on the fact that booze was illegal and make millions?

Yes, they did. But the gangs of that period were not born during prohibition. Most existed before the enactment of prohibition, and many had made the local saloons their places of operation. Gambling and prostitution were their big money-makers, and this was true even during prohibition. Although a great deal of smuggling and illegal selling of beverage alcohol did take place, the huge payoffs to police and politicians that were necessary for the gangs to continue their illegal business cut heavily into their profits.

There were a number of gangland killings because of infringement of territory or other rivalries. Probably the best known of these tragedies is the St. Valentine's Day Massacre in Chicago. Bugs Moran sent six men to Detroit with three trucks to bring back 150 cases of illegal booze that had been smuggled out of Canada. When the liquor runners returned to Chicago on February 14, 1929, they made their way to a garage on LaSalle Street, as they had been instructed to do. When they pulled into the garage, a police car pulled in behind them with machine guns firing. When the encounter was over, all of Bugs Moran's men were dead, and the trucks had been driven away by the police officers, who were not police officers at all but members of Al Capone's gang.

Newspapers across the nation played up this gangland ambush, labeling it the St. Valentine's Day Massacre and blaming the prohibition law for the killings.

Journalists romanticized each gangster encounter and with the increase in communications that took place during the 1920s it is not surprising that the public became convinced that a national crime wave was under way.

Actually, far more people die violent deaths today as a result of alcohol use than were victims of violence during prohibition.

In her book *Alcohol and Your Health*, Louise Burgess explains what the so-called prohibition crime wave meant to the average American:

Crime increased among the underworld—due to the gangster's own private war over huge booty made possible by the illegal trade. The general public may have lived the war vicariously, in the press and through radio, but remained almost entirely unharmed. Gang leaders, in fact, were wary of upsetting the status quo. Enforcement officers, federal and local, thus could more easily look the other way. Also, as one bootlegger said: "It just don't make no sense to rough up our customers."[10]

While there seems to have been an increase in crime toward the end of the prohibition era, an examination of what happened after repeal makes it clear that prohibition was not the cause. In 1935, the director of federal prisons announced that the prison population had increased by twenty-five percent in the first year of repeal. He said:

 . . . the increase in liquor crimes . . . and practically all kinds of crime, have carried us beyond the estimates; the relief we were expecting to get through the repeal of prohibition did not materialize.[11]

The repeal of prohibition did not reduce crime, and the years since then have seen a continual increase in violence and illegal activities. Although prohibition may not be the answer to America's crime problem, it is certain that we cannot drink our way to domestic tranquility.

Was the Great Depression brought on by prohibition?

The uninformed think so.

The truth is that prohibition introduced America to unparalleled prosperity, which continued for nearly a decade. The real story of economic progress during prohibition is seldom told.

Those working for national prohibition had predicted that the millions of dollars being spent on booze would be turned to better uses. They envisioned better home life, an increase in industrial output, and rising living standards. And they were right.

In his book *The Amazing Story of Repeal*, Fletcher Dobyns writes:

> The economic aspect of prohibition was made the subject of exhaustive investigation by many men who were honest, competent, and disinterested. They found that prohibition brought about a great decrease in the amount of liquor consumed, increased the dependability and efficiency of labor, reduced industrial accidents and losses; that the loss to the farmers was more than offset by the increased sale of milk, bread, vegetables, meat, and other farm products; that the closing of the saloons had increased the value of adjacent real estate; that it greatly increased the power of the people to purchase necessary and useful articles and stimulated every line of legitimate industry; and that it increased the national wealth and promoted the prosperity of all classes not directly or indirectly interested in the liquor business.[12]

Prohibition plugged the hole in the nation's pocketbook. Wages that had been supporting liquor producers were used to purchase manufactured goods. Samuel Crowther, an economist of that era, explained the impact on the nation's economy when money was used to purchase merchandise rather than beverage alcohol. He wrote:

> We shall entirely disregard, for the moment, any possible effects of the liquor on [the purchaser] and think of it only as a way of spending money. But a purchaser of liquor sets in motion a very small chain of purchasing, while, if the family of the man has that twenty dollars to spend, they will put it out into goods

which require a deal of labor and start many chains of purchasing. Or the same effect will be had if they save part of the money. . . .

There is an absolute unanimity of opinion that the wage earners are spending more on their families than they ever did and that the standards of living are constantly growing higher. . . . Prohibition, it appears from the letters which I have received, has definitely switched the spending of wages for drink to the spending of wages for goods. These letters in themselves present a really remarkable record—and it is an unprejudiced, first-hand record having to do only with the effects of spending on prosperity. All of the writers are in a position to know what they are talking about. . . . The answers from everywhere in the country are the same—the working men are spending little or nothing for drink and a great deal on their families. . . .

By the rerouting of at least two-thirds of the money which formerly went for drink into the buying of useful goods, a higher level of general living has been established in this country. The higher level has brought higher wages and still higher levels of living.

We have as a nation been infinitely more prosperous since prohibition than ever before. We are definitely going forward. It would seem that prohibition is fundamental to our prosperity—that it is the greatest blow which has ever struck poverty.[13]

Leaders in all areas of life—many of whom had opposed prohibition—spoke out in its favor, including industrialists, social workers, labor leaders, doctors, and journalists.

But the people spoke the loudest. In 1928 they elected the driest Congress in history.

Why, then, was prohibition finally repealed?

Three factors brought about the repeal of the Eighteenth Amendment:

1) The efforts of the Association Against the Prohibition Amendment (AAPA).

2) The problem of enforcement.

3) The Great Depression.

In his book *Booze, Bucks, Bamboozle, and You!* Ross J. McClennan writes:

> The "masses" brought prohibition to the nation but a small group of millionaires were the initial organizers of a concerted effort to destroy the law of the land which had been approved by the voters.[14]

McClennan is referring to the AAPA, an organization born in April of 1919. The group's goal was to make prohibition inoperative. Revealing the tactics of the organization, McClennan writes:

> The personal contributions of this group of millionaires was only a small part of the support they threw behind the Association Against the Prohibition Amendment. Through their interlocking directorates they controlled over forty billion dollars of invested funds. With this power structure at their command they could determine the policy of a great many powerful newspapers and magazines. By determining the editorial policy of newspapers and magazines the AAPA could brainwash the readers of such publications. And this they did.[15]

What was the reason for the persistent push by AAPA members? For many, it was the hope that tax revenues from the sale of alcohol would eliminate the need for income tax as a means of financing the federal government. We all know that didn't work.

Enforcement of prohibition was a serious problem. To know this one needs only to examine newspaper accounts and arrest records of the prohibition era. The enforcement of a dry law by wet politicians and police departments carries with it inherent difficulties.

Al Capone complained that his payoffs to politicians and policemen totaled approximately $30 million a year. Estimates of amounts of money changing hands illegally to keep the booze flowing on a nationwide scale are extremely high. That should not surprise us, given the nature of man.

After a decade of prosperity under prohibition, the Roaring Twenties came to an end with a collapse of the stock market in October of 1929. To this day, most Americans associate the Great Depression with prohibition. Prohibition did not bring on the depression and in fact may have postponed it, but the depression's environment of panic allowed prohibition's opponents to capitalize on the ills of the nation and the people's distress and bring about repeal of the Eighteenth Amendment.

Author Dobyns begins his chapter on the depression as follows:

> We come now to the most dishonest and shameless phase of the entire wet propaganda—its capitalization of the depression.[16]

Dobyns is justified in making his strong statement. Propaganda laid the cause of the depression at prohibition's door. But nothing could have been further from the truth.

Thomas Nixon Carver, professor of political economics at Harvard University, answered the charges, saying:

> The depression is world-wide. The countries which

do not have prohibition are worse off than we are. Some
of them are practically bankrupt. They owe us money
which they cannot repay. England went off the gold
standard. Wages are now paid in a depreciated cur-
rency.... This was probably necessary in order to save
British industries from wholesale bankruptcy. It was
not prohibition that put Great Britain in such a position.

If prohibition produced the depression in this country, it worked as a very slow poison. Wartime restrictions on the liquor traffic were adopted in 1918 as
economic measures. Liquor was prohibited to soldiers
and sailors. Wartime prohibition came in 1919, practically without opposition. The Eighteenth Amendment
became operative in January 1920. For 10 years this
country enjoyed unexampled prosperity under prohibition. We were able to lend billions of dollars to other
countries after the war was over. This helped them to
rehabilitate their industries and set their men to work.
In spite of that they now claim that they were unable to
pay us back. In other words, they plead bankruptcy. We
may have to accept that plea and forgive their debts.
That would put us in the position of a rich creditor
forgiving a poor debtor. Yet every one of those countries
is spending enough on drink to more than pay what
they owe us.

It was not prohibition that put them in such a position. It was not prohibition that makes them such poor
customers for what we have to sell. It does not seem that
prohibition in this country could cause a world depression.[17]

Prohibition did not cause the depression, and repeal
did not end it. After repeal, the depression became
more serious. The production of booze did not bring
about the promised miraculous economic recovery.
Relief was years away.

On December 6, 1933, prohibition ended. And the drinking levels of Americans, which had been substantially reduced during fourteen dry years, began to rise again, bringing a trail of misery, crime, and death that continues to this day.

What lessons were learned from the prohibition experience?

At least the following:

1) Public indignation against evil can produce results.

2) A decrease in alcohol use on a nationwide scale is achievable.

3) When the use of beverage alcohol is decreased, the result is a higher quality of life for all.

Today, with alcohol use in America an epidemic, nearly all solutions to the problem center on treatment for alcoholics. But this does little to prevent others from becoming afflicted. No plague has ever been stayed by only treating the sick. Prevention is necessary.

What can be done to stem the rising tide of alcohol use in America?

In periods of revival, solutions to such problems have risen from the churches. What is the present spiritual climate in the land? Is there hope for a new awakening that will empty the taverns and fill the places of worship?

Do Christians always agree on the booze question? Should they?

What does the Bible say about beverage alcohol?

9

WINE IN THE OLD TESTAMENT

Christians are divided on the question of drinking.

In America, the majority of evangelical churches take at least a nominal stand against the use of alcohol. The move is on to water down this position. Social drinking is increasing, even among fundamentalists. Some avoid hard liquor but look upon the use of beer and wine as permissible. Many churches use fermented wine for their Communion services.

In Europe, a great number of professing Christians use wine or beer regularly under the guise of necessity because of "bad water" or "culture acceptance." A surprising number of traveling American churchgoers who are abstainers at home feel free to drink while in Europe. Even some missionaries drink where that is the custom, saying their convictions allow them to "do as the Romans do."

The whole issue is discussed by Mark A. Noll in his article, "America's Battle Against the Bottle," which appeared in *Christianity Today.*

Some evangelicals have made opinions on liquor

more important for fellowship and cooperation than attitudes toward the person of Christ or the nature of salvation. This is particularly unfortunate since the Bible speaks clearly about Christ and salvation, but not about the question of total abstinence.[1]

But is Noll's conclusion correct? Is there no biblical absolute on the alcohol question? Has God left us without direction as to what our position should be on one of the most important social and moral issues?

The Bible certainly is not silent about beverage alcohol. The word "wine" appears more than two hundred times in the King James version of the Old Testament. This word was translated from a number of different Hebrew words.

The first biblical record of intoxication has to do with Noah, whom the Bible calls a "preacher of righteousness." Commenting on Noah's drunkenness, F. B. Meyer has written:

> Noah's sin reminds us how weak are the best of men; liable to fall, even after the most marvelous deliverances. The love of drink will drag a preacher of righteousness into the dust. Let us see to it that we fall not into this temptation ourselves; and that we tempt not others.[2]

Some feel that Noah was unaware of the process of fermentation and that intoxicating wine was unknown before the Flood. Matthew Henry alludes to this, saying:

> The drunkenness of Noah is recorded in the Bible, with that fairness which is found only in Scripture, as a case and proof of human weakness and imperfection, even though he may have been surprised into the sin; and to show that the best of men cannot stand upright,

unless they depend upon Divine grace, and are upheld thereby.[3]

Commentators' efforts to excuse Noah might be wishful thinking. This man of faith was not the first nor the last to stumble after being greatly blessed by God. Interesting as it may be to speculate about conditions before and after the Flood, we simply do not know the facts. We do know, however, that the day after his intoxication, Noah cursed members of his family. That unpleasant result of drinking is still common today.

Fermented wine was used in Sodom. Lot's drunkenness after his deliverance from that city before its destruction testifies to the use of booze there (Gen. 19:30–38). Lot's drinking episode ended in immorality, a frequent companion of intoxication. How different the future for Lot's descendants might have been had he or his daughters not included wine among the provisions they hurriedly put together in their last-minute escape from doomed Sodom.

The prophet Daniel, an abstainer, tells of Belshazzar's (the king of Babylon) last night on earth. He says that on that fateful night the king made a great feast for a thousand of his lords. Wine flowed freely. As the drinking continued, Belshazzar called for the golden and silver vessels that Nebuchadnezzar had taken out of the Jerusalem temple, so that he and his friends could drink wine out of them. (Alcohol often encourages irreverence.) When the party reached its peak of impiety, God called the drinking to a halt and warned the king of coming judgment. Before morning, Belshazzar had lost his position, his kingdom, and his life. (See Dan. 5.)

In the Old Testament priests were instructed not to drink wine or any kind of strong drink.

And the LORD spake unto Aaron, saying, Do not drink wine nor strong drink, thou, nor thy sons with thee, when ye go into the tabernacle of the congregation, lest ye die: it shall be a statute forever throughout your generations: And that ye may put difference between holy and unholy, and between unclean and clean; And that ye may teach the children of Israel all the statutes which the LORD hath spoken unto them by the hand of Moses (Lev. 10:8–11).

Upon skimming these verses one might conclude that this command to abstain from all alcoholic beverages had only to do with serving in the tabernacle. A more thorough reading, however, with special attention given to verse 10, makes it clear that abstaining from beverage alcohol was to be a way of life for the priests. This lifestyle was to demonstrate the difference between holy and unholy, between clean and unclean. In this context, the use of intoxicating beverages is seen as unholy and unclean, and the aim of the priest's lifestyle was to set an example before the people.

Joseph Seiss, an outstanding Lutheran theologian of the nineteenth century, gives the following commentary on this text.

The history of strong drink is the history of ruin, of tears, of blood. It is, perhaps, the greatest curse that has ever scourged the earth. It is one of depravity's worst fruits—a giant demon of destruction. Men talk of earthquakes, storms, floods, conflagrations, famine, pestilence, despotism, and war; but intemperance in the use of intoxicating drinks has sent a volume of misery and woe into the stream of this world's history, more fearful and terrific than either of them. It is the Amazon and Mississippi among the rivers of wretchedness. It is the Alexander and Napoleon among the warriors upon the peace and good of man. It is like the pale horse of the Apocalypse whose rider is Death, and at whose heels follow hell and destruction. It is an evil which is limited to no age, no continent, no nation, no party, no sex, no period of life. It has taken the poor man at his toil and the rich man at his desk, the senator in the halls of state and the drayman on the street, the young man in his festivities and the old man in his repose, the priest at the altar and the layman in the pew, and plunged them together into a common ruin. It has raged equally in times of war and in times of peace, in periods of depression and in periods of prosperity, in republics and in monarchies, among the civilized and among the savage. Since the time that Noah came out of the ark, and planted vineyards, and drank of their wines, we read in all the histories of its terrible doings, and never once lose sight of its black and bloody tracks.[4]

No wonder the priests—the Lord's representatives—were commanded to refrain from drinking intoxicating wine. It is a destroyer of people, an enemy of those they were to lead in the way of life.

Samson's mother was commanded not to drink wine or strong drink while awaiting the birth of her child, because Samson was to be dedicated to God in a special way.

And the angel of the LORD appeared unto the woman, and said unto her, Behold now, thou art barren, and bearest not: but thou shalt conceive, and bear a son. Now therefore beware, I pray thee, and drink not wine nor strong drink, and eat not any unclean thing (Judg. 13:3,4).

Rulers were forbidden to use intoxicating wine.

It is not for kings, O Lemuel, it is not for kings to drink

wine; nor for princes strong drink: Lest they drink, and forget the law, and pervert the judgment of any of the afflicted (Prov. 31:4,5).

Solomon gave a blanket command, setting forth the biblical principle that all fermented wine is to be avoided.

Look not thou upon the wine when it is red, when it giveth his colour in the cup, when it moveth itself aright (Prov. 23:31).

The word *look* as Solomon used it means "to lust for" or "to desire." He is simply saying that we are to have nothing to do with wine after it has fermented.

There are many Old Testament warnings about the effects of intoxicating wine.

Wine is a mocker.

Wine is a mocker, strong drink is raging: and whosoever is deceived thereby is not wise (Prov. 20:1).

Heavy drinking brings poverty.

For the drunkard and the glutton shall come to poverty: and drowsiness shall clothe a man with rags (Prov. 23:21).

The use of intoxicating wine brings trouble physically and socially.

Who hath woe? who hath sorrow? who hath contentions? who hath babbling? who hath wounds without cause? who hath redness of eyes? They that tarry long at the wine; they that go to seek mixed wine (Prov. 23:29,30).

Intoxicating wine ultimately harms the user.

At the last it biteth like a serpent, and stingeth like an adder (Prov. 23:32).

Beverage alcohol is the companion of immorality and untruthfulness.

Thine eyes shall behold strange women, and thine heart shall utter perverse things (Prov. 23:33).

The urge to drink can be so strong that it overcomes good judgment, making one forget the misery of his last binge.

They have stricken me, shalt thou say, and I was not sick; they have beaten me, and I felt it not: when shall I awake? I will seek it yet again (Prov. 23:35).

When religious leaders indulge in strong drink, they deceive their followers as to the realities of life and the importance of getting right with God while there is time.

Come ye, say they, I will fetch wine, and we will fill ourselves with strong drink; and to morrow shall be as this day, and much more abundant (Is. 56:12).

Drinking makes a proud and selfish person.

Yea also, because he transgresseth by wine, he is a proud man, neither keepeth at home, who enlargeth his desire as hell, and is as death, and cannot be satisfied, but gathereth unto him all nations, and heapeth unto him all people (Hab. 2:5).

The description, then, of beverage alcohol as set

forth in the Bible is that of an enemy attacking its users and robbing them of everything that is good in life. Human experience bears this out. To quote the eloquent Seiss again on the evils of strong drink:

> Egypt, the source of science—Babylon, the wonder and glory of the world—Greece, the home of learning and of liberty—Rome with her Caesars, the mistress of the earth—each in its turn had its heart lacerated by this dreadful canker-worm, and thus became an easy prey to the destroyer. It has drained tears enough to make a sea, expended treasure enough to exhaust Golconda, shed blood enough to redden the waves of every ocean, and rung out wailing enough to make a chorus to the lamentations of the underworld. Some of the mightiest intellects, some of the most generous natures, some of the happiest homes, some of the noblest specimens of man, it has blighted and crushed, and buried in squalid wretchedness.[5]

In the Old Testament, as well as in the New, wine is often a symbol of God's judgment and wrath. In writing of God's chastening of His people, the psalmist says they have been made to drink the "wine of astonishment" (Ps. 60:3).

The wrath of God prepared for the wicked is pictured as a cup full of fermented wine.

> For in the hand of the LORD there is a cup, and the wine is red; it is full of mixture; and he poureth out of the same: but the dregs thereof, all the wicked of the earth shall wring them out, and drink them (Ps. 75:8).

The prophet Jeremiah saw God's fury symbolized in a cup of wine.

> For thus saith the LORD God of Israel unto me; Take

the wine cup of this fury at my hand, and cause all the
nations, to whom I send thee, to drink it. And they shall
drink, and be moved, and be mad, because of the sword
that I will send among them. Then took I the cup at the
LORD'S hand, and made all the nations to drink, unto
whom the LORD had sent me (Jer. 25:15–17).

In summary, then, the Old Testament records
specific tragedies resulting from the use of beverage
alcohol. It singles out special people and groups whose
lives were to be examples to others, and they are
commanded not to drink intoxicating beverages. Clear
Old Testament commands declare that we are not to
look upon fermented wine with longing nor desire.

Intoxicating wine mocks, impoverishes, affects
health, injures its users, and contributes to immorality and dishonesty. It warps character, encouraging
selfishness and greed. It is seen as a symbol of God's
wrath and judgment.

But there is another side to the question. Some Old
Testament verses speak of wine as a blessing, a symbol
of prosperity, a source of cheer and gladness. Consider
these examples:

Therefore God give thee of the dew of heaven, and the
fatness of the earth, and plenty of corn and wine (Gen.
27:28).

And the vine said unto them, Should I leave my wine,
which cheereth God and man, and go to be promoted
over the trees? (Judg. 9:13).

He causeth the grass to grow for the cattle, and herb
for the service of man: that he may bring forth food out
of the earth; And wine that maketh glad the heart of
man, and oil to make his face to shine, and bread which
strengtheneth man's heart (Ps. 104:14,15).

How can wine be both a curse and a blessing, a symbol of judgment and a symbol of prosperity? How shall we explain these seeming contradictions?

More than one hundred years ago, Dr. William Patton struggled with these same questions. Serving as a pastor in New York City in an era when alcohol use was rampant, Patton became concerned about the detrimental effect that beverage alcohol was having on people in his community. He was disturbed that many who were heavy drinkers quoted the Bible as a defense for their drinking. Let Patton tell his story:

> I soon found that the concession so generally made, even by ministers, that the Bible sanctions the use of intoxicating drinks, was the most impregnable citadel into which all drinkers, all apologists for drinking, and all venders of the article, fled. This compelled me, thus early, to study the Bible patiently and carefully, to know for myself its exact teachings. I collated every passage, and found that they would range under three heads: 1. Where wine was mentioned with nothing to denote its character; 2. Where it was spoken of as the cause of misery; and 3. Where it was mentioned as a blessing, with corn and bread and oil—as the emblem of spiritual mercies and of eternal happiness. These results deeply impressed me, and forced upon me the question, Must there not have been two kinds of wine? So novel to my mind was this thought, and finding no confirmation of it in the commentaries to which I had access, I did not feel at liberty to give much publicity to it—I held it therefore in abeyance, hoping for more light. More than thirty-five years since, when revising the study of Hebrew with Professor Seixas, an immiment [sic] Hebrew teacher, I submitted to him the collation of texts which I had made, with the request that he would give me his deliberate opinion. He took the manuscript and, a few days after, returned it with the statement, "Your discriminations are just; they

denote that there are two kinds of wine, and the Hebrew Scriptures justify this view." Thus fortified, I hesitated no longer, but, by sermons and addresses, made known my convictions.[6]

Patton's book, *Bible Wines or Laws of Fermentation and Wines of the Ancients*, has become a classic on the subject.

Having given himself to serious study of the Hebrew and Greek texts and their biblical contexts, Patton discovered the following surprising facts.

(1) The Hebrew words translated "wine" in the Bible do not always mean fermented or intoxicating wine.

(2) The Hebrew word *yayin*, most often translated "wine" in the Old Testament, means grape juice in any form—fermented or unfermented. The true meaning can only be determined by the text.

(3) The Hebrew word *tirosh*, also translated "wine," in all but one possible case means "new wine," "unfermented wine." This word was used repeatedly in the original text in the places where wine has a good textual connotation.

(4) Many wines of the ancients were boiled or filtered to prevent fermentation, and these were often considered the best wines.

So, light begins to break through. The Bible speaks of two kinds of wine: good wine and bad wine, unfermented wine and fermented wine, wine that does not intoxicate and wine that does intoxicate.

While a detailed study of words translated "wine" in the Bible is provided in Appendix A, consideration here of the two most frequently used words, *yayin* and *tirosh*, will be sufficient to put the question to rest.

Tirosh, translated "wine" in the Old Testament, means new wine or grape juice. It sometimes refers to the juice still in the grapes before pressing. Consider these examples:

> Therefore God give thee of the dew of heaven, and the fatness of the earth, and plenty of corn and wine [*tirosh*] (Gen. 27:28).

Note the association with corn, speaking of the harvest.

> All the best of the oil, and all the best of the wine, and of the wheat, the firstfruits of them which they shall offer unto the Lord, them have I given thee. And whatsoever is first ripe in the land, which they shall bring unto the Lord, shall be thine; every one that is clean in thine house shall eat of it (Num. 18:12,13).

The wine here (*tirosh*) is part of the offering of the firstfruits, that is, the earliest gatherings of the harvest. It is brought freshly pressed to the altar.

> And he will love thee, and bless thee, and multiply thee: he will also bless the fruit of thy womb, and the fruit of thy land, thy corn, and thy wine, and thine oil, the increase of thy kine, and the flocks of thy sheep, in the land which he sware unto thy fathers to give thee (Deut. 7:13).

The use of *tirosh* in this text is again with corn and oil—part of the harvest. The reference is unmistakably to new wine, grape juice.

> That I will give you the rain of your land in his due season, the first rain and the latter rain, that thou mayest gather in thy corn, and thy wine, and thine oil (Deut. 11:14).

Note the gathering of corn and wine in the harvest with an unmistakable reference to the wine (*tirosh*) being juice still in the grapes—unfermented wine.

> And the vine said unto them, Should I leave my wine [*tirosh*], which cheereth God and man, and go to be promoted over the trees? (Judg. 9:13).

This interesting verse is part of Jotham's parable, in which the trees call to the vine to come and reign over them. But the vine refuses because it does not want to leave its wine, which cheers God and man. There is no doubt that the wine, or grape juice, is still in the grapes. It is in this unfermented state that wine cheers God and man, because it is part of the blessed abundant harvest.

> And that we should bring the firstfruits of our dough, and our offerings, and the fruit of all manner of trees, of wine and of oil, unto the priests, to the chambers of the house of our God; and the tithes of our ground unto the Levites, that the same Levites might have the tithes in all the cities of our tillage (Neh. 10:37).

Again, wine (*tirosh*) is spoken of as part of the offering of the firstfruits. Fermentation would have been impossible. If any doubts remain, perhaps Nehemiah 10:39 will settle them.

> For the children of Israel and the children of Levi shall bring the offering of the corn, of the new wine, and the oil, unto the chambers, where are the vessels of the sanctuary, and the priests that minister, and the porters, and the singers: and we will not forsake the house of our God.

Isaiah further confirms Patton's findings.

> Thus saith the LORD, As the new wine is found in the cluster, and one saith, Destroy it not; for a blessing is in it: so will I do for my servants' sakes, that I may not destroy them all (Is. 65:8).

There is little wonder that Patton's study brought him to the conclusion that there are two kinds of wine—fermented and unfermented. In this passage the wine is described as still being in the cluster, and "a blessing is in it."

Now read Joel's thrilling prophecy of millenial blessings.

> And it shall come to pass in that day, that the mountains shall drop down new wine, and the hills shall flow with milk, and all the rivers of Judah shall flow with waters, and a fountain shall come forth of the house of the LORD, and shall water the valley of Shit-tim (Joel 3:18).

We shall drink wine in the kingdom—new wine that drops from the vines in the vineyards that grow on the mountains. This wine is unfermented; it is not intoxicating.

And we shall share it together when Christ reigns as King.

While the Hebrew word *tirosh* is translated "wine" 38 times, the word used for wine most often in the Old Testament is *yayin*, which appears 141 times. *Young's Analytical Concordance* defines *yayin* as "what is pressed out, grape juice." In his article, "Did Jesus Turn Water Into Intoxicating Wine?" Lloyd Button writes:

> It should be made clear that the English word "wine" used in the Bible is a translation of a number of words in

the Hebrew and Greek languages referring to various products of the vine. Some Bible dictionaries insist that the usual meaning of the word wine is fermented. "Yayin" in Hebrew and "oinos" in Greek are the general terms for wine, and as we note later in this article can refer to both fermented and unfermented wine.[7]

Patton quotes Prof. M. Stuart on the meaning of the word *yayin*.

In the Hebrew Scriptures the word *yayin*, in its broadest meaning, designates grape-juice, or the liquid which the fruit of the vine yields. This may be new or old, sweet or sour, fermented or unfermented, intoxicating or unintoxicating. The simple idea of grape-juice or vine-liquor is the basis and essence of the word, in whatever connection it may stand. The specific sense which we must often assign to the word arises not from the word itself, but from the connection in which it stands.[8]

Since the word *juice* appears only once in the Old Testament (in Song of Solomon 8:2, referring to the juice of a pomegranate), and since the geographical setting of the Scriptures is a land of vineyards where grape juice was plentiful, it is understandable that many of the biblical texts having to do with the fruit of the vine refer to grape juice in its unfermented state. This is often the case with the meaning of the word *yayin* (translated "wine").

Note how *yayin* speaks of unfermented wine in the following texts.

And Melchizedek king of Salem brought forth bread and wine [*yayin*]: and he was the priest of the most high God (Gen. 14:18).

Remembering that fermented wine was forbidden to the priests (Lev. 10:9,10), we conclude that the wine Melchizedek carried was unfermented grape juice— wine that does not intoxicate.

> He causeth the grass to grow for the cattle, and herb for the service of man: that he may bring forth food out of the earth; And wine that maketh glad the heart of man, and oil to make his face to shine, and bread which strengtheneth man's heart (Ps. 104:14,15).

Because of expressions that today are associated with the use of intoxicating beverages, the reader is likely to think that "making glad the heart of man" is similar to "feeling good" as a result of drinking wine. Nothing could be further from the truth. Fermented wine does not make a person glad; it simply induces sleep in some areas of his brain.

A reading of Psalm 4:7 shows that the source of gladness in one's heart may come from corn and wine, but that this gladness reaches its peak through appreciating God's goodness. In both texts, degrees of gladness are the result of appreciating God's provision. At its height, this gladness comes from understanding God's spiritual blessings. To a lesser degree, it is a result of God's care of man in giving the harvest of food, wine, and oil.

Jeremiah says:

> As for me, behold, I will dwell at Mizpah to serve the Chaldeans, which will come unto us: but ye, gather ye wine, and summer fruits, and oil, and put them in your vessels, and dwell in your cities that ye have taken (Jer. 40:10).

This scene of the harvest calls for wine (*yayin*) to be

brought in from the fields; it is to be gathered with the summer fruits. Therefore, it would be unfermented and nonintoxicating.

In his masterful work *The Use of "Wine" in the Old Testament*, Dr. Robert P. Teachout, associate professor of Old Testament at Detroit Baptist Divinity School, concludes his lengthy and detailed study of *yayin* in this way.

> Therefore, the unified idea which is inherent in the word *yayin* is not that of a "fermented wine" per se (with divine approval dependent upon an assumed restriction of the quantity ingested, an assumption which is not explicit anywhere in the Old Testament). Instead, the comprehensive idea which the word conveys is that of a "grape beverage" (with the implied fermentation or its lack to be determined objectively only from the divine approval or disapproval of the beverage indicated by any context).[9]

In Teachout's judgment, *yayin* is intended to mean "grape juice" seventy-one times and "fermented wine" seventy times in the Old Testament.

Symbolically, intoxicating wine speaks of judgment and wrath. In contrast, nonintoxicating wine speaks of spiritual blessings.

> Ho, every one that thirsteth, come ye to the waters, and he that hath no money; come ye, buy, and eat; yea, come, buy wine and milk without money and without price (Is. 55:1).

William Patton says:

> In all the passages where good wine is named there is no lisp of warning, no intimations of danger, no hint of disapprobation, but always a decided approval.

How bold and strongly marked is the contrast:
The one the cause of intoxication, of violence, and of woes.
The other the occasion of comfort and peace.
The one the cause of irreligion and of self-destruction.
The other the devout offering of piety on the altar of God.
The one the symbol of divine wrath.
The other the symbol of spiritual blessings.
The one the emblem of eternal damnation.
The other the emblem of eternal salvation.[10]

The supposed riddle of the use of wine in the Old Testament is no longer a mystery. And it is encouraging to know that one does not have to be a master of the original languages to determine the type of wine spoken of in each text. Charles Wesley Ewing explains:

> . . . if a reader will just consider the context surrounding the word he can easily understand whether the fermented or unfermented grape juice was intended. Wherever the use of wine is prohibited or discouraged it means the fermented wine. Where its use is encouraged and is spoken of as something for our good it means the unfermented.[11]

But what about the New Testament?
Didn't Jesus make wine?
Was fermented wine used in the first Communion service?
Does the New Testament sanction the use of beverage alcohol?

10

JESUS AND WINE

"You Too Can Turn Water Into Wine," declared an advertisement for home winemaking sets in a popular American magazine. But the slow process of home winemaking is far from the instantaneous transformation of water to wine brought about by our Lord's first miracle. And the end product would be far from that served to the guests at the wedding in Cana.

Did Jesus ever make, use, or approve of the use of intoxicating wine? This question cannot be answered by appealing to the Greek language (see Appendices). *Oinos*, the Greek word most often translated "wine" in the New Testament, refers to grape juice in all its forms—unfermented and fermented, nonintoxicating and intoxicating. In the Septuagint Version, *oinos* is used to translate both *yayin* and *tirosh* from the Hebrew text. We have already seen that these two words mean grape juice in any form, with *tirosh* particularly referring to new wine or wine of the harvest.

In the New Testament, then, as well as in the Old, an understanding of the word *wine* can only come through studying the contexts.

Jesus is associated with wine in the following New Testament settings.

(1) The parable of the wine and the wineskins (Matt. 9:17; Mark 2:22; Luke 5:37–39).

(2) Jesus' statement that He had come eating and drinking, and the public reaction to that statement (Matt. 11:19; Luke 7:33,34).

(3) Jesus making wine at the wedding in Cana (John 2:3–10).

(4) The Lord's Supper instituted (Matt. 26:27–29; Mark 14:23–25; Luke 22:17–20).

(5) Wine offered to Jesus on the cross (Mark 15:23).

We will consider each of these biblical settings, in each case using the text that gives the greatest detail.

The wine and the wineskins

> And no man putteth new wine into old bottles; else the new wine will burst the bottles, and be spilled, and the bottles shall perish. But new wine must be put into new bottles; and both are preserved. No man also having drunk old wine straightway desireth new: for he saith, The old is better (Luke 5:37–39).

The primary message of this parable does not concern wine, but salvation. The law was to be fulfilled, and the age of grace was to begin. John the Baptist had announced, "For the law was given by Moses, but grace and truth came by Jesus Christ" (John 1:17).

Jesus was telling his hearers that law and grace were not to be mixed, and that salvation by grace would not be enhanced by adding legalism. One would be born again and receive eternal life entirely on the basis of faith. This wonderful message must not be considered simply an addition to the law; that would be like putting new wine into old bottles. G. H. Lang writes:

If the Christian who has known this heavenly liberty and gladness returns to legality and externalism, even though it be copied from that of the Mosaic economy, he will presently give a display of the Lord's prediction in this passage. The worn-out system will break to pieces and he himself will lose his joy and strength.[1]

In the case of wine, putting new wine into old bottles would hurry the fermentation and thus burst the bottles or wineskins. The aim was evidently to keep the new, unfermented wines sweet and nonalcoholic as long as possible. Old bottles would contain residues of yeast and would cause the wine to ferment quickly. William Patton comments:

> The new bottles or skins being clean and perfectly free from all ferment, were essential for preserving the fresh unfermented juice, not that their strength might resist the force of fermentation, but, being clean and free from fermenting matter, and closely tied and sealed, so as to exclude the air, the wine would be preserved in the same state in which it was put into those skins.
> Columella, who lived in the days of the Apostles, in his recipe for keeping the wine "always sweet," expressly directs that the newest must be put in a "new amphora," or jar.[2]

So then, the gospel must be kept pure, not mixed with legalism. But what about Luke 5:39? Does not Jesus say that the old wine is better?

Not at all. He simply says that one *who has been drinking* old wine says it is better. This shows the Lord's understanding of the habit-forming effect of beverage alcohol. His statement stands true today. Try to sell grape juice on skid row and you will probably have no takers. Those who drink old wine (intoxicating wine) prefer it. They are hooked on it.

The point of this parable is that the new wine (salvation) is better than the old wine (legalism). Jesus accurately predicted the reaction of his hearers. The Pharisees chose their legalism rather than the new life Christ offered them. Dr. H. A. Ironside explains:

> And so these Pharisees would go away saying, "we are satisfied with the old wine," and legalists and worldlings are like that today. They are apparently content with what they are trying to enjoy down here and do not care what God offers them in Christ Jesus.[3]

The secondary message of this parable, then, actually argues for the superiority of new (unfermented) wine, using it as a picture of salvation.

Jesus eating and drinking, and the crowd's reaction

> For John the Baptist came neither eating bread nor drinking wine; and ye say, He hath a devil. The Son of man is come eating and drinking; and ye say, Behold a gluttonous man, and a winebibber, a friend of publicans and sinners! But wisdom is justified of all her children (Luke 7:33–35).

Jesus contrasts Himself to John the Baptist, who was a Nazarite. Jesus was a Nazarene because He came from Nazareth, but He was not a Nazarite. Nazarites were people who had been called to live under a special vow, a totally different lifestyle, in order to show their total devotion to God. Among other things, they were to eat nothing made from grapes, including grape juice.

And the LORD spake unto Moses, saying, Speak unto the children of Israel, and say unto them, When either man or woman shall separate themselves to vow a vow of a Nazarite, to separate themselves unto the LORD: He shall separate himself from wine and strong drink, and shall drink no vinegar of wine, or vinegar of strong drink, neither shall he drink any liquor of grapes, nor eat moist grapes, or dried. All the days of his separation shall he eat nothing that is made of the vine tree, from the kernels even to the husk (Num. 6:1–4).

Jesus was not restricted by a Nazarite's vow. Therefore, He came eating grapes and drinking the fruit of the vine, all of which was called wine. This brought a reaction from His enemies. They called Him a gluttonous man and a winebibber (a tippler or heavy drinker).

But it is a mistake to accept the word of Jesus' enemies as truth. On two other occasions they said He had a devil (John 7:20 and John 8:48). Crowd reaction is certainly not safe ground for building sound doctrine. Jesus answered their charge by saying that God's wisdom is shown to be true by all who accept it. His righteous life would prove their accusations false.

Jesus making wine at the marriage feast in Cana

And the third day there was a marriage in Cana of Galilee; and the mother of Jesus was there: And both Jesus was called, and his disciples, to the marriage. And when they wanted wine, the mother of Jesus saith unto him, They have no wine. Jesus saith unto her, Woman, what have I to do with thee? mine hour is not yet come. His mother saith unto the servants, Whatsoever he saith unto you, do it. And there were set there six waterpots of stone, after the manner of the purifying of the Jews, containing two or three firkins apiece. Jesus

saith unto them, Fill the waterpots with water. And
they filled them up to the brim. And he saith unto them,
Draw out now, and bear unto the governor of the feast.
And they bare it. When the ruler of the feast had tasted
the water that was made wine, and knew not whence it
was: (but the servants which drew the water knew;) the
governor of the feast called the bridegroom, And saith
unto him, Every man at the beginning doth set forth
good wine; and when men have well drunk, then that
which is worse: but thou hast kept the good wine until
now. This beginning of miracles did Jesus in Cana of
Galilee, and manifested forth his glory; and his disciples believed on him (John 2:1–11).

The events surrounding Jesus' first miracle need
little explanation. He had been invited to a wedding
and had been accompanied by His disciples. During
the feast following the ceremony, the host ran out of
wine. Mary presented the problem to Jesus, and He
provided wine enough for all the guests.

There are two important facts to keep in mind.

(1) The wine Jesus made did not conform to
twentieth-century standards; it conformed only to
the standards of that day. To those at the feast, all
juice of the grape, fermented or unfermented, was
considered wine.

(2) Whatever Jesus made that day was consistent
with His character.

Jesus had come to fulfill the Scriptures. Scores of
Old Testament prophecies were fulfilled in His birth,
life, death, and resurrection. The gospel declares this
truth.

For I delivered unto you first of all that which I also
received, how that Christ died for our sins according to
the scriptures; And that he was buried, and that he rose

again the third day according to the scriptures (1 Cor. 15:3,4).

We have already seen that in the Old Testament, fermented wine symbolized wrath and judgment. Its use was prohibited. It is inconceivable, then, that Jesus would have violated this biblical principle by making more than 120 gallons of intoxicating wine to be served to the wedding guests.

In 1907, Dr. R. A. Torrey wrote:

> The wine provided for the marriage festivities at Cana failed. A cloud was about to fall over the joy of what is properly a festive occasion. Jesus came to the rescue. He provided wine, but there is not a hint that the wine He made was intoxicating. It was fresh-made wine. New-made wine is never intoxicating. It is not intoxicating until sometime after the process of fermentation has set in. Fermentation is a process of decay. There is not a hint that our Lord produced alcohol, which is a product of decay or death. He produced a living wine uncontaminated by fermentation.[4]

One of the great Bible scholars of this century, Dr. William Pettingill, wrote:

> I do not pretend to know the nature of the wine furnished by our Lord at the wedding of Cana, but I am satisfied that there was little resemblance in it to the thing described in the Scriptures of God as biting like a serpent and stinging like an adder (Prov. 23:29–32). Doubtless rather it was like the heavenly fruit of the vine that He will drink new with His own in His Father's kingdom (Matt. 26:29). No wonder the governor of the wedding feast at Cana pronounced it the best wine kept until the last. Never before had he tasted such wine, and never did he taste it again.[5]

Isn't that characteristic of our Lord? For believers, the best is yet to come!

Those who base their use of beverage alcohol on the miracle at Cana are on shaky ground. The excellent commentary *Barnes On the New Testament* places the burden where it belongs.

> No man should adduce this instance in favour of drinking wine unless he can prove that the wine made in the "water pots" of Cana was just like the wine which he proposes to drink.[6]

The water pots of Cana were filled with *kingdom wine* and the earthly supply is exhausted, not to be replenished until the King returns.

The Lord's Supper

> And as they were eating, Jesus took bread, and blessed it, and brake it, and gave it to the disciples, and said, Take, eat; this is my body. And he took the cup, and gave thanks, and gave it to them, saying, Drink ye all of it; For this is my blood of the new testament, which is shed for many for the remission of sins. But I say unto you, I will not drink henceforth of this fruit of the vine, until that day when I drink it new with you in my Father's kingdom (Matt. 26:26–29).

Those who conclude that fermented wine was used in the first Communion service do so without biblical support. Jesus spoke only of "the cup" and "the fruit of the vine."

Is fermented wine the fruit of the vine? Charles Wesley Ewing argues:

> Fermented wine is not a product of the vine. Chemi-

cally it is entirely different from the sweet and unfermented grape juice. Fermented wine is 14% alcohol, and it has other constituents that are not found in the fresh grape juice. Alcohol does not grow on the vine. It is not a vine product. Alcohol is the product of decay, the product of fermentation. It is produced by the process of spoiling.[7]

The fruit of the vine used in the first Communion service speaks of the blood of Christ. Moses called unfermented grape juice the "pure blood of the grape" (Deut. 32:14). Decay has not taken place in fresh grape juice, and that fact is vital in the symbol of the blood of Christ. David prophesied that the body of Christ would be totally preserved from decay or corruption.

For thou wilt not leave my soul in hell; neither wilt thou suffer thine Holy One to see corruption (Ps. 16:10).

Peter says this is a reference to the resurrection of Christ (Acts 2:31). So we are given the assurance both prophetically and by the New Testament Scriptures that the body of Christ experienced no corruption or decay. It would be improper, then, for a product of decay (fermented wine) to be used to symbolize His blood. Commenting on this text in his book *The King of the Jews*, Dr. John R. Rice says:

The cup the disciples drank at the Lord's supper is nowhere called wine, but "the fruit of the vine." We believe it was simply grape juice. Even if the word wine had been used, wine in the Bible means grape juice, whether fermented or unfermented. Fermented wine, with microbes of decay, would not picture the perfect blood of a sinless Christ.[8]

William Patton concurs, stating:

> Leaven, because it was corruption, was forbidden as an offering to God . . . If leaven was not allowed with the sacrifices, which were the types of the atoning blood of Christ, how much more would it be a violation of the commandment to allow leaven, or that which was fermented, to be the symbol of the blood of atonement? We cannot imagine that our Lord, in disregard of so positive a command, would admit leaven into the element which was to perpetuate the memory of the sacrifice of himself, of which all the other sacrifices were but types.[9]

Fermented wine would have been out of place at the Lord's Supper for a number of reasons. Perhaps the most important has to do with the holy character of this experience. Remember that in the Old Testament, the priests were forbidden to use wine. Jesus is our great High Priest, the fulfiller of the Scriptures. We can be sure that He remained consistent when establishing the Communion service and therefore did not use intoxicating wine as the symbol of His blood.

The final statement of our Lord on the Communion service settles the issue. All Christians who take Communion are to do so in anticipation of the coming kingdom. We have already seen that the wine of the kingdom is unfermented. If we are to look forward to the wine of Eden and Cana during the kingdom, it would be inconsistent to use intoxicating wine when remembering the death of our Savior and King.

Wine offered to Jesus on the cross

> And they gave him to drink wine mingled with myrrh: but he received it not (Mark 15:23).

The wine offered to Jesus at the time of His crucifixion was, without doubt, intoxicating wine. Its purpose was to make the pain more bearable.

In His most trying hour, Jesus refused intoxicating wine.

And so should we.

11

WINE IN THE CHURCH

The early church had scarcely begun its task of world evangelism when its members were accused of drunkenness. On the day of Pentecost the believers were given the gift of tongues, which enabled them to preach the gospel in all the languages of the people gathered at Jerusalem for the feast. As a result of this miracle, thousands were converted to Christ. Others, confused by hearing the many languages, began to mock the disciples, saying: "These men are full of new wine" (Acts 2:13).

In his book *Lectures on Acts*, H. A. Ironside shares an interesting experience similar to that of the disciples on the day of Pentecost.

This situation was illustrated very clearly to me some years ago in San Francisco when a group of us were in the habit of going down to the worst part of the city every Saturday night where hundreds of sailors from the ships in the harbor would pass. We held a street meeting from eight o'clock until midnight, speaking to all classes of men. One speaker, now a missionary in the

Argentine Republic, was a Spaniard by birth, yet spoke fluently French, Italian, Portuguese and other languages. When he would see a group of French seamen passing (the name of their ship upon their caps), he would suddenly call out to them in their own language and speak to them for perhaps twenty minutes; and then, as he sighted a group of Portuguese sailors (easily distinguished by their uniforms) he would swing over and talk to them in Portuguese and they would gather in close. Later he might speak to a group of Spaniards or Mexicans and then perhaps to some Italians. There was rarely a Saturday night when he did not speak in all these different languages. More than once I have seen persons come up and say, "What is the use of listening? He is drunk. You can't understand a word he says!" They did not know the language, and that is the way it was on Pentecost. Peter and his companions were not acting strangely—that wasn't the point; but as they spoke in different languages, those who couldn't comprehend came at once to the conclusion that they were drunk.[1]

There is, however, another dimension to the mockery of the crowd on that day. They accused the disciples of being drunk on new wine (*gleukos*), a product that does not intoxicate. Patton writes:

To account for the strange fact that unlettered Galileans, without previous study, could speak a multitude of languages, the mockers implied they were drunk, and that it was caused by new wine (*gleukos*). Here are two improbabilities. The first is that drinking alcoholic wine would teach men languages. We know that such wines make men talkative and garrulous; and we also know that their talk is very silly and offensive. In all the ages, and with the intensest desire to discover a royal way to knowledge, no one but these mockers has hit upon alcohol as an immediate and successful teacher of languages.

The second improbability is, that *gleukos*, new wine, would intoxicate.[2]

The crowd's charge that the disciples are drunk on new wine seems to be mockery run wild. In effect, they are saying, "These abstainers are drunk on grape juice." Of course, they lacked understanding concerning the miracle that was taking place that day. The promise of the Father had been fulfilled. The Holy Spirit had come. Believers had been baptized into the body of Christ and filled with the Holy Spirit. The church had been born.

Here, as in other portions of the Bible, it is not safe to build doctrines or convictions on the mockery of a crowd. The charge of intoxication on the day of Pentecost was false.

Heavy drinking did become a problem in the church at Corinth, however, even degrading the Communion service. Paul rebuked these carnal Corinthian believers, saying:

When ye come together therefore into one place, this is not to eat the Lord's supper. For in eating every one taketh before other his own supper: and one is hungry, and another is drunken (1 Cor. 11:20,21).

The breaking of bread in the church at Corinth had deteriorated into gluttony and drunkenness. Does this mean the Christians in Corinth used fermented wine for their Communion service?

Yes, without question.

But their practice had developed apart from apostolic instruction. In celebrating Communion, they had moved from the fruit of the vine to intoxicating wine. There is no evidence that this is true in any of the other local churches spoken of in the New Testament. In Paul's instruction concerning the Communion service, given to correct the errors at Corinth, he avoids the use of the word "wine," describing the first Communion as follows:

> For I have received of the Lord that which also I delivered unto you, That the Lord Jesus the same night in which he was betrayed took bread: And when he had given thanks, he brake it, and said, Take, eat: this is my body, which is broken for you: this do in remembrance of me. After the same manner also he took the cup, when he had supped, saying, This cup [containing the fruit of the vine—see Matt. 26:29] is the new testament in my blood: this do ye, as oft as ye drink it, in remembrance of me. For as often as ye eat this bread, and drink this cup, ye do shew the Lord's death till he come (1 Cor. 11:23–26).

Note Paul's reference to the "cup," and his reminder that the Communion service looks forward to the Lord's return, when Christians will share kingdom wine with their Savior.

Carnality was rampant in the church at Corinth, and its problems went far deeper than just the use of intoxicating wine (although that is often a companion of trouble). Believers there were given to divisions, gossip, and confusion. Immorality was common. Their church services were disorderly. Paul had to remind them that God is not the Author of confusion (1 Cor. 14:33). Perhaps most sobering is the revelation that some of their number had become sick and others had died because of their irreverence at the Communion table.

The effects of beverage alcohol have not changed with the passing of centuries. Commenting on the judgment brought upon Nadab and Abihu because of their irreverence in the service of God (Lev. 10), J. A. Seiss wrote:

> If the effects of alcoholic stimulation went no further than to cloud the mind and stupefy the natural senses of those who indulge in it, it would not be so bad. The great mischief is that, as it clouds the moral nature, it kindles all the bad passions into redoubled activity. It not only enfeebles and expels all impulses of good, but it quickens and enthrones every latent evil, and fits a man for the ready performance of any vile and sacrilegious deed.[3]

Wine brought confusion and chastening to members of the Corinthian church. It will bring the same to all who follow in their steps.

On two occasions Paul spoke favorably of temperance. Three times he emphasized the importance of being temperate. Once he urged moderation. Does this mean Paul favored the controlled or limited use of beverage alcohol?

Let us consider the texts in question.

> And as he reasoned of righteousness, temperance, and judgment to come, Felix trembled, and answered, Go thy way for this time; when I have a convenient season, I will call for thee (Acts 24:25).

> But the fruit of the Spirit is love, joy, peace, longsuffering, gentleness, goodness, faith, Meekness, temperance: against such there is no law (Gal. 5:22,23).

In both of the above settings, "temperance" means self-control. In the first reference, the apostle is rebuking Felix, his judge, because of the governor's lack of self-control, warning him of the consequences of his loose living.

As part of the fruit of the Spirit, temperance (self-control) gives evidence of a life that is totally yielded to God and under the direction of His Spirit.

When Paul writes of being temperate, he is speaking of the disciplined Christian life. Here are examples of his use of that word.

> And every man that striveth for the mastery is temperate in all things. Now they do it to obtain a corruptible crown; but we an incorruptible (1 Cor. 9:25).

> For a bishop must be blameless, as the steward of God; not selfwilled, not soon angry, not given to wine, no striker, not given to filthy lucre; But a lover of hospitality, a lover of good men, sober, just, holy, temperate (Titus 1:7,8).

> But speak thou the things which become sound doctrine: That the aged men be sober, grave, temperate, sound in faith, in charity, in patience (Titus 2:1,2).

In every case appearing in the Bible, "temperance" refers to self-control and living a disciplined life.

The word "moderation" appears once in the Bible in Philippians 4:5: "Let your moderation be known unto all men. The Lord is at hand."

A careful reading of this text reveals that Paul is not calling for moderation in drinking. The word translated *moderation* here means gentleness and has to do with our attitude toward others in view of Christ's return.

But doesn't Paul say we should drink to the glory of God? "Whether therefore ye eat, or drink, or whatsoever ye do, do all to the glory of God" (1 Cor. 10:31).

There is no question about his command to drink to the glory of God, but there is not a hint that this drinking involves beverage alcohol. Disobedience does not bring glory to God. One can only act to the glory of God when his action is within the framework of biblical revelation.

We have already seen that intoxicating wine is presented in the Bible as an enemy, a mocker, a producer of poverty, and a symbol of divine wrath. Therefore, it is inconceivable that Paul would urge his readers to use this destructive substance in the hope of bringing glory to God.

Some may question why God allows fermentation if He forbids the use of fermented drinks. The question may have been answered best by Clarence True Wilson in a debate with Clarence Darrow on the subject of prohibition.

"I bought some grape juice and put it away for a month and God turned it into wine," Darrow said.

Wilson replied, "How about eggs? Nature and time will do the same thing to them, but I don't insist on eating them."

Darrow had no answer.[4]

Those given to the use of fermented wine were not allowed to hold office in the New Testament church. Among the qualifications for the office of a bishop was a restriction concerning the use of wine.

> A bishop then must be blameless, the husband of one wife, vigilant, sober, of good behaviour, given to hospitality, apt to teach; Not given to wine, no striker, not greedy of filthy lucre; but patient, not a brawler, not covetous (1 Tim. 3:2,3).
>
> Likewise must the deacons be grave, not double-tongued, not given to much wine, not greedy of filthy lucre (1 Tim. 3:8).

Some have concluded that the "much wine" of this text allows for the use of some wine by deacons. But Paul is simply saying that there are to be no drinking deacons. He does not open the door in this text to *some* wine anymore than to *some* gossip or *some* greed. Had he done so, his counsel for deacons would have placed him in direct conflict with the entire body of Old Testament teaching on the subject. The only valid reason for the use of a little wine is for medicinal purposes (1 Tim. 5:23). In Appendix K of Robert Teachout's thesis, he explains.

In the light of the conclusions drawn earlier that there is no explicit Old Testament justification for assuming that wine drinking is ever appropriate for the saint, even in moderation, it is important to indicate briefly that the New Testament evidence concurs with, or at least is not contrary to, this conclusion. The reason that this appendix is necessary is that a superficial understanding of 1 Timothy 3:8 might lead one to the belief that the New Testament qualifications for leadership stated there, "not given to much wine" (AV), requires a re-evaluation of the Old Testament evidence. However . . . this verse too can be readily harmonized with the remainder of Scripture which leads to the position that the Bible always condemns the use of intoxicating beverages in any amount.[5]

We can be certain that Paul would not have compromised the unity of the Scriptures in order to provide a few cups of wine for thirsty deacons in the early church.

The message that comes through concerning the servants of God—from Old Testament priests to New Testament pastors and deacons—is that God requires total abstinence. Patton says:

That both Paul and Timothy understood that total abstinence was an essential qualification for the Christian pastor is evident from the compliance of Timothy. In the same letter v.23, Paul advises Timothy, "Drink no longer water, but use a little wine for thy stomach's sake and thine oft infirmities." The fact is plain that Timothy, in strict accordance with the direction, "not given to wine," that is, not with or near wine, was a total abstainer. The recommendation to "use a little wine" is exceptional, and strictly medicinal.[6]

Those who choose to use intoxicating beverages sometimes hold up Paul's medical prescription for Timothy's stomach trouble as justification for their drinking. They seem to forget that this single instruction to use wine was strictly for a medicinal purpose. Neither can we be sure that the wine prescribed was fermented. Charles Wesley Ewing offers this interesting possibility:

Timothy was a native of the city of Lystra in Lycaonia. At the time of Paul's letter to him, Timothy was at Ephesus. Both of the places are in Asia Minor. The water in that region was strongly alkali and was upsetting to Timothy's stomach. Paul was giving him advice on how to get rid of his stomach disorder. It was the practice in those days, and it is still practiced today in Syria, Mesopotamia, and other parts of Asia Minor, that when people drank this alkali water they would mix it with a spoon full of jam made from boiling the juice of grapes that are similar to our own Concord grape. The acids in the grape jam would neutralize the alkali in the water and make it fit for the stomach. This is what Paul was telling Timothy to do.[7]

Actually, we cannot be sure whether the wine prescribed by Paul for Timothy's stomach trouble was fermented or unfermented. In either case, the text provides no encouragement for the use of fermented wine except for the sick. This position is consistent with the one set forth in the Old Testament, and therefore, it is the one we can safely conclude to be correct.

In the finest hour of the church, beverage alcohol was shunned by earnest and dedicated believers. This also has been true during periods of great revival in church history. Today, claiming Christian liberty, many believers choose to imbibe alcoholic drinks. But Christian liberty was never intended to open the door to the destroyer.

We are free from the law, but all Christian freedom is within the boundaries of love. Liberty extends to that which builds up others, not that which causes weaker ones to stumble. The use of beverage alcohol by church members today is a tragic compromise of biblical standards that can only weaken the church and disillusion new converts. Unless there is a reversal of present trends, the church will become powerless through misusing the liberty intended to be her strength.

Alcohol enslaves. It is not a champion of liberty. And the church should be the one place on earth where thirsty souls can find peace apart from the presence of mood-altering drugs.

12

BETTER THAN WINE

... thy love is better than wine (Song of Solomon 1:2).

The Bible forbids the use of wine as we know it today.

All wine?

Every drop.

Don't even long for it.

Look not thou upon the wine when it is red, when it giveth his colour in the cup, when it moveth itself aright (Prov. 23:31).

Medicinal use is the only exception (1 Tim. 5:23).

Three studies have confirmed our conclusions: a study of wine in the Old Testament, a study of Jesus and wine, and a study of wine in the church. Three sacred sources speak for total abstinence from intoxicants. And as Solomon wrote, "A threefold cord is not quickly broken" (Eccl. 4:12).

The person who abstains from alcoholic beverages is on solid biblical ground. This position is not popular. Nevertheless, we are convinced it is the scriptural one.

But why does God withhold beverage alcohol from His children?

The answer may be found in Psalms 84:11, ". . . no good thing will he withhold from them that walk uprightly."

Alcoholic beverages are not good for us. Therefore, our heavenly Father forbids their use.

And there is more.

We are called to the abundant life.

> The thief cometh not, but for to steal, and to kill, and to destroy: I am come that they might have life, and that they might have it more abundantly (John 10:10).

While it is clear that God has commanded us to abstain from the use of beverage alcohol, we can be sure that it is not His purpose to empty our lives of happiness. When our Lord removes something from a life, He always replaces it with something better. And many of His good gifts are "better than wine."

God's love is better than wine.

In London, an alcoholic thought by many to be hopeless was placed under the care of a psychiatrist, but received little help. An evangelistic crusade was being held in the city, and he was invited to attend. There he heard of God's love for him, and he responded to the invitation to receive Christ as his Savior.

When the new convert was about to fall asleep that night, he reached for his bottle to take his customary last drink of the day but found himself unable to continue his old habit. Getting out of bed, he emptied the bottle of liquor down the sink drain. When he awakened in the morning he reached, by force of habit, for his usual morning "bracer." It was not there. But he was not disappointed. He knew that he had been set free.

Why? What power had changed his life?

The power of God. And God's power is made available to those who respond to His love.

But there is more to the story.

The former alcoholic phoned his psychiatrist and told him that he had lost a patient. Describing his experience at the crusade, he said, "I am now a new man."

"Sounds fine!" the psychiatrist replied. "Maybe I can find help where you found it."

The troubled counselor began to attend the evangelistic services and was also moved by the message of God's love. Like the patient he had just lost, he opened his heart to Christ.

A few days later, at a large hotel in London, there was a gathering of fifty converts of the crusade. There, both the doctor and his former patient told of the changes that had taken place in their lives as a result of accepting the gift of God's love.[1]

The list of former drinkers who have become total abstainers through responding to God's love is long. Names known to thousands—like Mel Trotter, Billy Sunday, and Oscar Van Impe (my own father)—come quickly to mind, but a great company of others have also testified to never drinking another drop of booze after receiving Christ as Savior.

Does this mean their lives became dull and meaningless because beverage alcohol was missing? Did they feel deprived? Did their obedience to the Bible and to the Holy Spirit on this matter put a damper on their enthusiasm for life? Was their abstinence a cross to bear?

Not at all.

Read their stories. Hear their dynamic testimonies. When one is born again, God's love more than compensates for the loss of liquor. The new convert has received something better than wine.

Christian fellowship is better than wine.

It is no accident that Paul's command, "Be not drunk with wine" (Eph. 5:18), is followed by instruction concerning Christian fellowship and family relationships.

> And be not drunk with wine, wherein is excess; but be filled with the Spirit; Speaking to yourselves in psalms and hymns and spiritual songs, singing and making melody in your heart to the Lord; Giving thanks always for all things unto God and the Father in the name of our Lord Jesus Christ; Submitting yourselves one to another in the fear of God (Eph. 5:18–21).

Christians are not to be controlled by wine but by the Holy Spirit. A life under the control of the Holy Spirit receives a new appreciation for other members of the family of God and for the members of his own family.

Some shrink from abstinence from beverage alcohol because they enjoy the fellowship of the bar, the association with drinking companions. There are crowds at taverns almost everywhere.

But God has something better . . . the fellowship of the family of God.

In her poem, "Fellowship," Wava Campbell writes:

> There is a special fellowship
> The world can never know;
> There is a special bond of love—
> At least I've found it so—
> Among the children of the Lord
> Wherever they are found;
> It started in the heart of God,
> And flows the world around.

> There is a special closeness
> That the world can't seem to feel;
> There is a special oneness
> That is very, very real;
> It's found in folks called "Christians"
> Wherever they may be;
> It started in the heart of God,
> And flows through you and me.[2]

When one considers the purpose for gathering, there is good reason for Christian fellowship to be superior to the fellowship of the bar. As drinking continues, the brain is dulled and participants are brought more and more under alcohol's control. This lowers restraint and self-control, bringing out the negative traits in each person's character. Man's sinful nature produces the "works of the flesh," described in the Bible as follows:

> Now the works of the flesh are manifest, which are these; Adultery, fornication, uncleanness, lasciviousness, Idolatry, witchcraft, hatred, variance, emulations, wrath, strife, seditions, heresies, Envyings, murders, drunkenness, revellings, and such like: of the which I tell you before, as I have also told you in time past, that they which do such things shall not inherit the kingdom of God (Gal. 5:19–21).

In contrast, Christian fellowship is an experience in building up one another. The sharing of spiritual truth in an atmosphere of thanksgiving and praise encourages submission to the Lord, which in turn increases the fruit of the Spirit in each life. The fruit of the Spirit is positive and powerful. Here are the qualities seen in the life under the control of the Holy Spirit.

But the fruit of the Spirit is love, joy, peace, longsuf-
fering, gentleness, goodness, faith, Meekness, temperance: against such there is no law (Gal. 5:22,23).

The motivation for Christian fellowship is love.
The motivation for the fellowship of the bar is lust.
Christian fellowship has often been a great source of
strength to those throwing off the chains of alcohol.
The tie that binds our hearts in Christian love is
superior to the bond of the bottle.

Family relationships are better than wine.
Continuing his comparison of intoxication from
wine to the filling of the Holy Spirit, Paul said:

Wives, submit yourselves unto your own husbands,
as unto the Lord. For the husband is the head of the
wife, even as Christ is the head of the church: and he is
the saviour of the body. Therefore as the church is
subject unto Christ, so let the wives be to their own
husbands in every thing. Husbands, love your wives,
even as Christ also loved the church, and gave himself
for it (Eph. 5:22–25).

God has ordained the home. The love of a devoted
wife, a faithful husband, and children is far more
valuable than alcohol's kick. Tragically, some have
only come to realize this after bringing other family
members through years of needless suffering.
In his book *The Old Lighthouse*, James R. Adair tells
of a period of false hope in the lives of Mel and Lottie
Trotter.

The Trotters moved to the little farm. For three glori-
ous months they were like honeymooners—with the
little boy. During that time Mel didn't touch the bottle

once. But what Lottie didn't know was that, for the last two months, a craving for drink was gnawing right through Mel's will power.

"You ta-take the baby inside, Honey," he told Lottie after a happy buggy ride one winter night. "I'll put the horse away, and I'll be right in."

But as the door closed, he wheeled the horse and buggy around and lashed the horse into a mad dash for town. Mel licked his lips, hardly able to wait to walk into the town saloon to order his first drink since he had moved to the farm.

After a few drinks, he was the same old wisecracking, devil-may-care Mel. "Hey fellows, drinks are on me. The bartender can have my horse and buggy. Let's drink 'im dry!"

Next morning an alcohol-soaked Mel staggered the 11 miles home, the 11 miles that was supposed to separate him and drink forever!

Wet-eyed and unsmiling, Lottie helped him in as tenderly as ever. Just her look and touch broke Mel up, and once again he swore off drinking and promised her he'd never do this to her again.[3]

But Lottie's heartaches were far from over. Mel's promises weren't worth much in those days. They would both cry buckets of tears as a result of his continued drinking.

Finally Mel came to Christ at the Pacific Garden Mission in Chicago, on his way to Lake Michigan to end it all. Of Trotter's conversion, Adair says:

He cried out to God that night for forgiveness, and God heard his sincere cry—for a sense of forgiveness flooded his soul.

From that night the thirst for alcohol left him, and God gave him complete victory. He learned that Jesus had said, "If any man thirst, let him come unto me, and drink" (John 7:37). Mel did just that, and Jesus took away the other evil thirst. Lottie's prayers were answered.[4]

Mel Trotter became an exemplary husband after his conversion. He used his natural talents in the service of Christ, founding sixty rescue missions in cities across America to reach those who had reached bottom as he had done.

One who is right with God has a deep appreciation for members of his family. The family circle is preferred to the drinking crowd. He is intoxicated with love for his wife. A dedicated Christian wife feels the same way about her husband. They are devoted to their children and consider them gifts from God. These blessings are enough to keep them thrilled with life, and they need no artificial high, no drug to counterfeit happiness.

One writer says that upon entering a businessman's office, she saw a placard nailed to his desk. It said:

WHICH?
Wife or Whiskey?
The Babes Or The Bottles?
HEAVEN OR HELL?

The businessman explained that he had printed the placard himself in order to keep his priorities right and to give proper attention to those he loved. The simple sign helped him remember that there are better things in life than booze.[5]

One of the best gifts God has given us is the love of our families. We must be careful not to waste the many opportunities we are given to be considerate of them. The satisfaction that comes from treating family members lovingly and from receiving love in return is better than wine.

God's daily blessings are better than wine.

Well-known author Upton Sinclair, in *The Cup of Fury,* tells of a number of people whose lives were shortened or ruined through the use of alcohol. Many wondered why Sinclair never became interested in drinking—even socially. His answer was that he had so many other kinds of intoxication—such as looking at nature, reading great poetry, listening to music, and seeking knowledge—that he never had the slightest interest in liquor.[6]

Sinclair's point is well taken. All of creation speaks of God's glory, "The heavens declare the glory of God; and the firmament sheweth his handiwork" (Ps. 19:1).

Our highest purpose should also be to glorify God.

His blessings are seen on every hand. We ought to be so intoxicated with God's goodness and with our opportunity to glorify Him that no fake high should ever be needed or desired. Let us avoid the error of Eden where our first parents focused on the one restriction given and thereby forfeited the abundance of the rest of the Garden.

Our heavenly Father has provided His best for us in every area of life. And His blessings are better than wine.

13

ENDING ALCOHOL'S LONG NIGHT

Everyone who drinks has an alcohol problem.

Each drink of beverage alcohol dulls the senses of the user and in some measure causes him to miss out on the thrill of being alive. Thomas Edison said:

> I do not drink alcoholic liquors. I have better use for my head. To put alcohol in the human brain is like putting sand in the bearings of an engine.[1]

Even moderate use of this drug carries with it the threat of dependency. Nearly all alcoholism begins with moderate or social drinking. Imbibing any amount brings the drinker under alcohol's power to some degree. Consider, then, Paul's declaration of independence: ". . . I will not be brought under the power of any" (1 Cor. 6:12).

The great majority of those who drink are periodically hazardous to others. Most alcohol-related highway accidents are not caused by alcoholics but by moderate to heavy drinkers. Many industrial mishaps are the result of careless acts by people who drink but do not consider themselves problem drinkers.

Who can know the boundary beyond which alcohol use becomes a health hazard? We know that drinking contributes to liver problems and that it adversely affects the brain and other organs, but who can say which drink is the straw that breaks the camel's back?

How many drinks does it take to trigger cancer of the esophagus?

How much booze can the larynx and pharynx tolerate before surrendering to cancer's attack?

According to the June, 1978, report of the Secretary of Health, Education, and Welfare, alcohol is involved in causing cancer in all these areas, but nobody knows just how much it takes to start trouble in each individual case. And who knows his own body's tolerance level?

Some people have advanced in their drinking to the point of alcoholism, a condition that ravages its victims, adversely affects family members, and costs society as a whole.

How does one detect alcoholism?

The U.S. Department of Health, Education, and Welfare says that any one of the following warning signals may indicate a drinking problem:

- Family or social problems caused by drinking
- Job or financial difficulties related to drinking
- Loss of a consistent ability to control drinking
- "Blackouts" or the inability to remember what happened while drinking
- Distressing physical and/or psychological reactions if you try to stop drinking
- A need to drink increasing amounts of alcohol to get the desired effect
- Marked changes in behavior or personality when drinking

- Getting drunk frequently
- Injuring yourself—or someone else—while intoxicated
- Breaking the law while intoxicated
- Starting the day with a drink[2]

In his book *The Trouble With Alcohol*, Tom Shipp lists three sets of questions that he has used in his work with alcoholics to determine whether drinkers are in the initial, the intermediate, or the final stages of alcoholism.

Here is the first set:

1. Do I have an intense personal reason for drinking? In other words, is my reason for drinking something other than social?
2. Am I experiencing a meaningful change from the use of alcohol? Do I drink to relieve tension, fears, anxieties, or inhibitions?
3. Do I find myself involved increasingly in thought about alcohol? Am I thinking about the problem of supply when I should be thinking about other things?
4. Are most of my friends heavy drinkers?
5. Has my drinking become more secretive, more guarded?
6. Am I drinking more often and more heavily than in the past? Am I kidding myself that by drinking beer and wine I am cutting down? Do I tell myself I am handling my problem because I maintain periods of not drinking at all in between alcoholic bouts?
7. When I start drinking, do I end up drinking more than I intended to drink? Do I find drunkenness occurring at closer intervals?
8. Have I failed to remember what occurred during

a drinking period last night, yesterday, or even a longer period ago?

9. Do I feel guilty, defensive, or angry when some-one wants to talk to me about my drinking?
10. Am I sneaking my drinks?
11. Have I stopped sipping my drinks and instead find myself gulping or tossing them down quickly?
12. Do I lie about my drinking?

Those who do not deal with their drinking problem will, of course, progress to a more serious condition. Consider Shipp's questions for drinkers who have reached the second stage of alcoholism.

1. Am I greatly concerned about protecting my supply? Have I developed an alibi system? Do I rationalize, or resent and suspect others? Do I find it necessary to lie to my employer, relatives, or friends to hide my drinking?
2. Have I lost control? Can I really stop drinking once I start? Do I find I need a drink to get over a drink? Do I want to drink alone? Have I become antisocial in my behavior?
3. Do I feel that other people are watching me?
4. Am I unduly critical of others such as my spouse, children, or the people with whom I work? Do I find a scapegoat rather than a solution to my problems? Is there always someone or something that I can blame for my drinking too much?
5. Has my behavior caused me to lose friends, fam-ily, and/or jobs?
6. Are people afraid of me while I am drinking?
7. Has another person—spouse, friend, fellow worker, or anyone else—said something to me about my drinking behavior?

Shipp says that at this intermediate stage, drinkers have not yet lost their health, their jobs, their families, their self-respect, or their self-confidence. They still can make decisions. For those in the final stage, he asks the following:

1. Do I really have a choice as to whether or not I can drink? No matter what my determination to stop, do I seem completely powerless over alcohol?
2. Do I drink for days or weeks at a time?
3. When I try to get off alcohol, do I develop the "shakes"?
4. Have my fears become worse? Am I more hostile toward others? Am I constantly afraid of something, but I do not know what?
5. Have I run out of excuses for what I am doing? Are there no more scapegoats around, no one to blame?[3]

Shipp's questions are revealing. They show that the alcoholic road leads ever downward into deeper bondage. One look at what these searching questions point to ought to bring any drinker up short. Still, many plunge more deeply into alcohol's pit every day.

How serious is alcoholism from a biblical standpoint?

Know ye not that the unrighteous shall not inherit the kingdom of God? Be not deceived: neither fornicators, nor idolaters, nor adulterers, nor effeminate, nor abusers of themselves with mankind, Nor thieves, nor covetous, nor drunkards, nor revilers, nor extortioners, shall inherit the kingdom of God (1 Cor. 6:9,10).

Many look upon alcoholism as a disease, but Jack

Odell, an alcoholic for seventeen years, offers a better analysis of the problem. In his book *Here's How!* he says:

Looking at alcoholism as just a disease and no more, there is no known cure.

Right there is where my doubts begin.

It used to take about a quart of whiskey a day to keep me going. Now I don't need it. I know personally a couple of dozen other people who used to be in the same boat. Many of them were in far worse shape than I. Now, and in some cases for many years, they just don't need it. I know of the case-histories of literally hundreds of others. Same thing. No bottle.

Even more surprising, all of these people are keeping dry (really I think they're being kept dry) without any effort. They don't have to grit their teeth and set their jaws. They don't use iron willpower to get past the saloons. They don't promise themselves just twenty-four hours more of sobriety. These people are dry simply because they like it that way. They don't refrain from drinking; they just don't need to drink.

They've been changed.

Transformed.

Saved.

My belief is that alcoholism is one of the symptoms of a far deeper and more serious disease. There are plenty of other symptoms too: bad temper, cruelty, greed, hypocrisy, compulsive promiscuity, fear, and self-pity—just to name a few. We've all known "dry alcoholics," people who go on periodic emotional binges without resorting to the jug.

The underlying disease is sin. These are the outbreaking manifestations of it. As spots go with measles, boils with blood impurities, and certain rashes with emotional stress, these things go with sin.

If this is true, we can only have a real cure when the

basic disease has been cured. Bleaching the spots doesn't cure the measles. I believe that all those people I mentioned who don't need to drink anymore have been cured of their real underlying disease, which was sin. Inevitably, the symptom—alcoholism—went away.[4]

Here, then, is hope for alcoholics: They are not more lost than other sinners, nor are they hopeless incurables. There is a known remedy for sin. As soon as the alcoholic sees his alcoholism as a symptom of this deeper sickness (sin) for which there is a cure, he can have hope.

A ray of light breaks through the darkness.

God loves alcoholics and other sinners.

When Jesus shed His blood on the cross for sinners, alcoholics were included along with the rest of us in God's great plan of redemption. The prophet Isaiah explained it as follows:

All we like sheep have gone astray; we have turned every one to his own way; and the LORD hath laid on him the iniquity of us all (Is. 53:6).

We are all sinners.

For all have sinned, and come short of the glory of God (Rom. 3:23).

And Christ died for all.

But God commendeth his love toward us, in that, while we were yet sinners, Christ died for us (Rom. 5:8).

In the death of Christ on the cross, there was payment for all the sins of every alcoholic: the broken promises, the profanity, the neglect of family members, and all the rest. That sacrifice also paid for all the sins of those who never touch a drop: the self-righteousness, the gossip, the bitterness, and other sins that alcoholics do not have a monopoly on.

The death and resurrection of Christ had been foretold by the prophets. He died, was buried, and rose again according to the Scriptures. This was God's plan of redemption, making peace with God possible.

Peace. What a good word!

But how does this theological truth provide peace for troubled people?

Through faith: "Therefore being justified by faith, we have peace with God through our Lord Jesus Christ" (Rom. 5:1).

Upon concluding a legal transaction in St. Louis, a Christian businessman said to his lawyer, "I have often wanted to ask you a question, but I have been a coward."

"Why?" asked the lawyer. "I didn't think you were afraid of anything. What is the question?"

"Why are you not a Christian?" asked the businessman.

The lawyer, a heavy drinker, looked downward and replied, "Is there not something in the Bible that says no drunkard shall have any part in the kingdom of God? You know my weakness."

"That is not my question," answered the Christian client. "I am asking you why you are not a Christian."

"Well," replied the lawyer, "I cannot recall that anyone ever asked me if I were a Christian, and I am sure nobody ever told me how to become one."

The Christian then drew close to the lawyer, read him some portions of the Bible, and said, "Let us get down and pray."

The lawyer prayed a simple prayer expressing his new faith in Christ and requesting that the power of alcohol would be broken in his life.

Later, speaking of his deliverance from alcohol, the converted lawyer said, "Put it down big, put it down plain, that God broke that power instantly." The lawyer's name was C. I. Scofield, known now to millions as the editor of the Scofield Reference Bible. His faith in Christ freed him from the enslaving power of alcohol.[5]

What happens when one places his faith in Christ as his Savior and Lord?

He is accepted by the Lord.

> He hath made us accepted in the beloved (Eph. 1:6).

He is totally forgiven.

> In whom we have redemption through his blood, the forgiveness of sins, according to the riches of his grace (Eph. 1:7).

He is justified (made just in the sight of God).

> Being justified freely by his grace through the redemption that is in Christ Jesus (Rom. 3:24).

He becomes a child of God.

> But as many as received him, to them gave he power to become the sons of God, even to them that believe on his name (John 1:12).

His body becomes the temple of God.

> Know ye not that your body is the temple of the Holy

Ghost which is in you, which ye have of God, and ye are not your own? For ye are bought with a price: therefore glorify God in your body, and in your spirit, which are God's (1 Cor. 6:19,20).

He becomes a new creation, having received a spiritual rebirth.

Therefore if any man be in Christ, he is a new creature: old things are passed away; behold, all things are become new (2 Cor. 5:17).

Imagine! A new life!

And here's the clincher: God equips this child of His with power to overcome any temptation that comes his way.

There hath no temptation taken you but such as is common to man: but God is faithful, who will not suffer you to be tempted above that ye are able; but will with the temptation also make a way to escape, that ye may be able to bear it (1 Cor. 10:13).

Life finally has meaning.

Jack Odell, now free from alcohol addiction, writes:

Because you have made the Great Transaction, you can expect the Holy Spirit to begin releasing and using those untapped capacities of yours..

If there's music in you, you'll sing because you finally have something to sing about.

If there's a book in you, you'll write because you finally have something to write about.

If there's capacity for service in you, you'll begin serving because you have a Master to serve.

Love? What was only shallowly provocative and lustful will become lovable and loving.

If your gift is for hard work, you'll work as never before: and happily so, because your energies are released and you have an indwelling Reinforcer.

If your gift is for laughter, you'll stop laughing at the cruel things. Then your laugh will become warm and contagious, and other people will want to join you. . . .

You'll be alive and creative and fulfilled for the first time.

Full-filled—because you're Christ-filled.[6]

Christ ends alcohol's long night for those who come to Him in faith. For many, all desire for alcohol ends at the moment of their salvation. Others face a daily battle, with increasing strength in times of conflict because of the power of God within and the resources of prayer, Bible study, and Christian fellowship. God's way for each Christian is always best. And every Christian is equipped to win.

If you are an alcoholic, you can end the long night of addiction that oppresses you.

Jesus said, "Come unto me, all ye that labour and are heavy laden, and I will give you rest" (Matt. 11:28).

Accept His loving invitation. Come to Him in faith. Call upon Him.

Confess Him before others. Start attending church services regularly. Build a strong prayer life. Saturate your mind with the Bible.

And don't forget those who love you. Go to them and tell them the long night is over.

No more waiting and worrying.

No more staring out of dark windows.

No more empty promises.

Your faith in Christ has opened the door to a new life.

14

CONFRONTING THE ENEMY

Beverage alcohol is an enemy.

Wars and plagues have destroyed their thousands. Alcohol has destroyed its ten thousands.

In the United States, alcohol use has reached crisis proportions. When this happened once before, the people reacted to the continuing tragedy and voted for a national prohibition. Is the general public ready to act against the enemy again?

The Third World Congress of the International Commission for the Prevention of Alcoholism and Drug Dependency met in Acapulco, Mexico, on August 28, 1979. There, William Plymat, executive director of the American Council on Alcohol Problems, compared our present national mood to that of an alcoholic who has "hit bottom." Pointing out the drinker's progression—lift (or euphoria) to relaxation to tension to hitting bottom—he said:

> During the period when national prohibition was in effect in the United States, consumption of alcohol was low and there were relatively few alcohol problems,

and the cost to society of these problems was very little. Then as legal alcoholic beverages came into being, the first results noticed by society seemed to be the tax revenues received by federal and state governments. This was what might be thought of as a "first" effect of legal alcohol.

As the years rolled by, the consumption of alcohol constantly increased. It was a gradually escalating condition, and in recent years the escalation has speeded up until today we are facing a monstrous problem. As a society, are we close to "hitting bottom?"[1]

Are the American people ready for another prohibition crusade?

Probably not.

Although prohibition brought many benefits to our nation, enforcement of a national nondrinking law is difficult in a society where drinking is accepted among a high percentage of those charged with running the government and keeping the peace. A perusal of newspapers published during the prohibition era bears this out.

So what can be done?

The most effective weapon against the enemy has always been a revived church. Revivals empty taverns and fill churches. And in our day, a widespread revival is desperately needed.

But how does one go about starting a revival? The evangelist Gypsy Smith was once asked that question. He answered, "Go home, lock yourself in your room, and kneel down in the middle of the floor. Draw a chalk mark all around yourself and ask God to start a revival inside that chalk mark. When He has answered your prayer, the revival will be on."[2]

America's greatest revival came about largely as the

result of one man's concern. A.C. Lanphier was working as a lay missionary in one of the crowded areas of New York City. He often became discouraged but drew strength from personal prayer. The thought occurred to him that others might find help for their needs by joining him in prayer. He let it be known that he was starting a series of weekly noon-hour prayer meetings, the first of which was held September 23, 1857.

For the first half hour, Lanphier prayed alone. Then, one by one, others came, until a total of six were praying. The next week twenty appeared, and the third week brought forty. By spring, more than twenty daily noon prayer meetings were occurring in New York City. Some of the largest churches were crowded to capacity. The police and fire departments opened their buildings for prayer services. Revival had begun. It spread across the land.

In a Boston meeting, a man said: "I am from Omaha, Nebraska. On my journey east I found a continuous prayer meeting all the way."[3] Lanphier's call to prayer had set his nation afire for God.

In times of revival, the spiritual weapons of prayer and evangelism have a powerful effect on society. The apostle Paul explained, "For the weapons of our warfare are not carnal, but mighty through God to the pulling down of strong holds" (2 Cor. 10:4).

These spiritual weapons have transformed societies before. During the Welsh revival, the following report appeared in the *Methodist Times*.

> In Cardiff, police reports show that drunkenness has diminished over 60 per cent, whilst on Saturday last the Mayor was presented by the Chief Constable with a pair of white gloves, there being no case at all on the charge sheet—an unprecedented fact for the last day of the year.

The same thing happened at the Swansea County Court on the previous Saturday, and the magistrate said, "In all the years I've been sitting here I've never seen anything like it, and I attribute this happy state of things entirely to the revival."

The streets of Aberdare on Christmas Eve were almost entirely free from drunkenness, and on Christmas Day there were no prisoners at all in the cells. At Abercarn Police Court, responsible for a population of 21,000, there was not a single summons on Thursday—a thing unknown since the court was formed fourteen years ago—and here, too, was enacted the ceremony of the white gloves.[4]

The evangelistic thrust that comes with genuine revival produces a devastating effect on the liquor business. Thousands converted during periods of evangelistic outreach have been weaned from the bottle for life. This is, without doubt, the most important and effective method available to us for solving the alcohol problem.

There are a number of other positive steps that can be taken. In implementing them, it is important to remember that the alcoholism rate in any population is directly related to the total consumption of alcohol in that community. That is, a decrease in the sale and use of beverage alcohol will lower the amount of alcoholism in any area.

In 1971, the Addiction Research Foundation of Ontario issued a report entitled "Alcohol Use and Alcoholism," based on research by Dr. Wolfgang Schmidt and Jan de Lint, members of the foundation staff. The report contained the following statement.

Alcoholism prevalence in a population is intimately related to the overall level of alcohol consumption. It follows, therefore, that any factor that affects the volume of consumption inevitably affects the alcoholism prevalence rate and vice-versa. . . .

Therefore, we have been forced to realize that there is no hope of reducing the numbers of alcoholics or of those who drink at levels hazardous to their health—without rolling back the overall consumption of alcohol throughout our society.[5]

Any move, then, that reduces the sale of beverage alcohol cuts down on alcoholism. The citizen action proposed below would help accomplish that goal.

Warning Labels

Recently, Sen. Strom Thurmond of South Carolina introduced a bill requiring warning labels on alcoholic beverages with twenty-four percent or more alcohol content. Thurmond's bill was attached to a funding bill for a government alcohol abuse program and passed the Senate by a vote of 68–21. It seems unlikely at this writing, however, that this kind of legislation will pass in the House of Representatives.

Reports say that most mail favored a widening of the labeling restriction to cover all alcoholic beverages. Two of the suggested label warnings were: *Caution: Consumption of alcoholic beverages may be hazardous to your health* and *Dangerous drug, illegal for minors.*

While labeling would not solve the nation's alcohol problem, it would identify beverage alcohol for what it is: *a dangerous drug and a health hazard.* Every move toward warning labels for alcoholic beverages should be supported.

Increased Taxation

Beverage alcohol does not pay its way. The public

cost that results from alcohol use exceeds alcohol revenues by more than four to one. This places every taxpayer in the position of subsidizing the alcohol industry—even citizens who have strong convictions against drinking. This deplorable situation should be corrected by increasing taxes on alcoholic beverages to a rate that covers the expenses incurred by their use.

Studies in a number of nations have demonstrated that the price of beverage alcohol affects the volume of consumption. When the price increases, drinking decreases. Cheap booze contributes to a lower quality of life. Increased taxation would bring double benefits: Revenue would increase because of the higher tax rate, and alcohol use would decrease because of the higher prices on the product made necessary by the rise in taxes. This action is long overdue, and there should be a grass-roots movement to bring it about.

Ban on advertising

Considering alcohol's destructive influence, it is mind boggling that this drug can be promoted publicly. Cigarette advertising is prohibited from television, but booze commercials are legion. And beverage alcohol is more deadly than tobacco. The social ills connected with drinking are far more hazardous than those associated with smoking.

Tackling the multi-million dollar commerce in alcohol advertising is admittedly a gigantic undertaking. Any movement entering this arena might justifiably feel like young David coming up against Goliath. But it must be remembered that David was the winner of that confrontation.

And since the cause is just, a ban on liquor advertising in all the media is achievable. If Billy Sunday and company could shut off the booze spigot nationwide, the descendants of that courageous band ought to be able to outlaw billboards and television commercials that hawk the virtues of beverage alcohol.

Limiting Outlets

Alcohol use increases with availability. In rural Finland, the government changed from a policy of allowing liquor sales only in certain outlets to permitting purchases in a variety of stores. The result was a fifty percent increase in alcohol consumption within one year.

Beverage alcohol is too available; it's pushed along in too many grocery carts.

Imagine the benefits to our nation if drinking could be *curbed* by fifty percent through stiff distribution laws! The impact on America's health would be akin to the discovery of a cancer cure.

Creative action is needed here, including stricter licensing, limits on the number of outlets based on the population of each community, and the banning of all alcoholic beverages from grocery stores. Bold moves in limiting outlets will bring cries of economic hardship by businesses and the threat of higher prices on other items, but it is time to get our priorities right. When lives are at stake, dollars must yield to sense.

Shortened Hours

Since the repeal of prohibition, liquor sellers have continued to whittle away at legal restrictions on hours that beverage alcohol may be sold. This aggressive policy has kept booze moving to users in an ever-increasing flow. The result has been expanded alcoholism and soaring public expenses related to the natural fallout from heavy drinking.

Reverse whittling is needed, and that is where the trouble begins.

Dollars are involved. Opposition is sure. What businessman wants to reduce his sales?

Nevertheless, the hard truth remains that a reduction of the sales of beverage alcohol must become the public goal if we are to cut down on alcoholism and its misery. Shortening the hours that alcoholic beverages can be sold is one way to move toward that goal.

The battle against alcohol use must be fought with both hands. We must reach out to alcohol's victims with love and concern, sharing the life-changing gospel of Christ with them, while at the same time restraining man's beloved enemy through every legal means at our disposal. This twofold responsibility calls both for aggressive evangelism and for personal involvement in a righteous cause . . . a cause that has been neglected in our day.

Society has "hit bottom." The time to act is now.

Great battles have been won in the past.

Let us confront the enemy with confidence.

The power of God is not diminished.

And we've been retreating long enough.

APPENDIX A

Dr. William Patton's Word Studies

(from his book *Bible Wines or Laws of Fermentation and Wines of the Ancients*)

Generic Words

Professor M. Stuart, in his *Letter to Rev. Dr. Nott,* February 1, 1848, says, page 11: "There are in the Scriptures (Hebrew) but two *generic words* to designate such drinks as may be of an intoxicating nature when fermented and which are not so before fermentation. In the Hebrew Scriptures the word *yayin,* in its broadest meaning, designates *grape-juice,* or the *liquid which the fruit of the vine yields.* This may be new or old, sweet or sour, fermented or unfermented, intoxicating or unintoxicating. The simple idea of *grape-juice* or *vine-liquor* is the basis and essence of the word, in whatever connection it may stand. The specific sense which we must often assign to the word arises not from the word itself, but from the connection in which it stands."

He justifies this statement by various examples which illustrate the comprehensive character of the word.

In the London edition (1863) of President E. Nott's *Lectures,* with an introduction by Tayler Lewis, LL.D., Professor of Greek in Union College, and several appendices by F. R. Lees, he says: "*Yayin* is a generic term, and, when not restricted in its meaning by some word or circumstance, comprehends vinous beverage of every sort, however produced. It is, however, as we have seen, *often* restricted to the fruit of the vine in its natural and unintoxicating state" (p. 68).

Kitto's Cyclopedia, article Wine: "*Yayin* in Bible use is a very general term, including every species of wine made from grapes (*oinos ampelinos*), though in later ages it became extended in its application to wine made from other substances."

Rev. Dr. Murphy, Professor of Hebrew at Belfast, Ireland, says: "*Yayin* denotes all stages of the juice of the grape."

"*Yayin* (sometimes written *yin, yain,* or *ain*) stands for the expressed juice of the grape—the context sometimes indicating whether the juice had undergone or not the process of fermentation. It is mentioned one hundred and forty-one times"—*Bible Commentary*, Appendix B, p. 412.

SHAKAR, "the second, is of the like tenor," says Professor Stuart, page 14, but applies wholly to a different liquor. The Hebrew name is *shakar*, which is usually translated *strong drink* in the Old Testament and in the New. The mere English reader, of course, invariably gets from this translation a wrong idea of the real meaning of the original Hebrew. He attaches to it the idea which the English phrase now conveys among us, viz., that of a *strong, intoxicating drink*, like our *distilled* liquors. As to *distillation*, by which alcoholic liquors are now principally obtained, it was utterly unknown to the Hebrews, and, indeed to all the world in ancient times. ... The true original idea of *shakar* is a *liquor obtained from dates or other fruits* (grapes excepted), or barley, *millet*, etc., which were dried, or scorched. And a decoction of them was mixed with honey, aromatics, etc."

On page 15 he adds: "Both words are *generic*. The first means vinous liquor of any and every kind; the second means a corresponding liquor from dates and other fruits, or from several grains. Both of the liquors have in them the *saccharine principle;* and therefore they may become alcoholic. But both may be kept and used in an *unfermented* state; when, of course, no quantity that a man could drink of them would intoxicate him in any perceptible degree. ... The two words which I have thus endeavored to define are the *only two* in the Old Testament which are *generic*, and which have reference to the subject now in question."

"SHAKAR (sometimes written *shechar, shekar*) signifies 'sweet drink' expressed from fruits other than the grape, and drunk in an unfermented or fermented state. It occurs in the O.T. twenty-three times"—*Bible Commentary*, p. 418. *Kitto's Cyclopedia* says: "*Shakar* is a generic term, including palm-wine and other *saccharine* beverages, except those prepared from the vine." It is in this article defined "*sweet drink.*"

Dr. F. R. Lees, page xxxii of his Preliminary Dissertation to the *Bible Commentary*, says *shakar*, "saccharine drink," is related to the word for sugar in all the Indo-Germanic and Semitic languages, and is still applied throughout the East, from India to Abyssinia, to the palm sap, the *shaggery* made from it, to the date juice and syrup, as well as to sugar and to the fermented palm-wine. It has by usage grown into a generic term for "drinks," including fresh juices and inebriating liquors other than those coming from the grape. See under the heading, "Other Hebrew Words" for further illustrations.

TIROSH, in *Kitto's Cyclopedia*, is defined "vintage fruit." In *Bible Commentary*, p. 414: "*Tirosh* is a collective name for the natural produce of the vine." Again, *Bible Commentary*, p. xxiv: "*Tirosh* is not wine at all, but the fruit of the vineyard in its natural condition." A learned biblical scholar, in a volume on the wine question, published in London, 1841, holds that *tirosh* is not wine, but fruit. This doubtless may be its meaning in some passages, but in others it can only mean wine, as, for example, Prov. iii. 10: "Thy presses shall burst out with new wine" (*tirosh*); Isa.lxii. 8: "The sons of the stranger shall not drink thy new wine" (*tirosh*).

"On the whole, it seems to me quite clear," says Prof. Stuart, p. 28, "that *tirosh* is a species of wine, and not a genus, like *yayin*, which means *grape-juice* in any form, or of any quality, and in any state, and usually is made definite only by the context."

"*Tirosh* is connected with corn and the fruit of the olive and the orchard nineteen times; with corn alone, eleven times; with the vine, three times; and is otherwise named five times: in all, thirty-eight times. . . . It is translated in the Authorized Version twenty-six times by wine, eleven times by new wine (Neh. x. 39, xiii. 5, 12; Prov. iii. 10; Isa. xxiv. 7, lxv. 8; Hos. iv. 14, ix. 2; Joel i. 10; Hag. i. 11; Zach. ix. 17), and once (Micah vi. 15) by 'sweet wine,' where the margin has new wine"—*Bible Commentary*, p. 415.

So uniform is the good use of this word that there is but one doubtful exception (Hosea iv. 11): "Whoredom and wine (*yayin*), and new wine (*tirosh*), take away the heart." Here are three different things, each of which is charged with taking away the heart. As whoredom is not the same as *yayin*, so *yayin* is not the same as *tirosh*. If physical intoxication is not a necessary attribute of the first, then why is it of the third, especially when the second is adequate for intoxication? If *yayin* and *tirosh* each means intoxicating wine, then why use both? It would then read, whoredom and *yayin* (intoxicating wine) and *tirosh* (intoxicating wine) take away the heart, which is tautological. The three terms are symbolical.

Whoredom is a common designation of idolatry, which the context particularly names. This steals the heart from God as really as does literal whoredom.

Yayin may represent drunkenness or debased sensuality. This certainly takes away the heart.

Tirosh may represent luxury, and, in this application, dishonesty, as *tirosh* formed a portion of the tithes, rapacity in exaction, and perversion in their use, is fitly charged with taking away the heart.

Certain interpreters imagine that only alcoholic drinks take away the heart; but we know from the Bible that pride, ambition, worldly pleasures, fulness of bread, Ezek. xvi. 49, and other things, take away the heart.

G. H. Shanks, in his review of Dr. Laurie, says: "In vine-growing lands, grapes are to owners what wheat, corn, flax, etc., are to agriculturists, or what bales of cotton or bank-notes are to merchants. Do these never take away the heart of the possessor from God?"

Other Hebrew Words

We extract from Dr. F. R. Lees' Appendix B of *Biblical Commentary* the following, pp. 415–418:

KHEMER is a word descriptive of the foaming appearance of the juice of the grape newly expressed, or when undergoing fermentation. It occurs but nine times in all, including once a verb, and six times in its Chaldee form of *khamar* or *khamrah*. Deut. xxxii. 14; Ezra vi. 9, vii. 22; Ps. lxxv. 8; Isa. xxvii. 2; Dan. v. 1, 2, 4, 23.

Liebig says: "Vegetable juice in general becomes turbid when in contact with the air *before fermentation commences*"—*Chemistry of Agriculture*, 3rd edition. "Thus it appears, *foam* or *turbidness* (what the Hebrews called *khemer*, and applied to the foaming blood of the grape) is no proof of alcohol being present"—*Bible Commentary*, Prelim. xvi. note.

AHSIS (sometimes written *ausis, asic, osis*) is specially applied to the juice of newly trodden grapes or other fruit. It occurs five times. Cant. viii. 2; Isa. xlix. 26; Joel i. 5, iii. 18; Amos ix. 13.

SOVEH (sometimes written *sobe, sobhe*) denotes a luscious and probably boiled wine (Latin, *sapa*). It occurs three times. Isa. i. 22; Hosea iv. 18; Nahum i. 14.

"It is chiefly interesting as affording a link of connection between classical wines and those of Judea, through an obviously common name, being identical with the Greek *hepsema*, the Latin *sapa*, and the modern Italian and French *sabe*—boiled grape-juice. The inspissated wines, called *defrutum* and *syraeneum*, were, according to Pliny (xiv. 9), a species of it. The last name singularly suggests the instrument in which it was prepared—the *syr*, or caldron"—*Bible Commentary*, Prelim. xxiii.

MESEK (sometimes written *mesech*), literally, a mixture, is used with its related forms, *mezeg* and *mimsak*, to denote some liquid compounded of various ingredients. These words occur as nouns four times, and in a verbal shape five times. Ps. lxxv. 8; Prov. xxiii. 30; Cant. vii. 2; Isa. lxv. 11. The verbal forms occur in Prov. ix. 2, 5; also, in Ps. cii. 9; Isa. xix. 14.

ASHISHAH (sometimes written *eshishah*) signifies some kind of fruit-cake, probably cake of pressed grapes or raisins. It occurs four times, and in each case is associated by the Authorized Version with some kind of drink. 2 Sam. vi. 19; I Chron. xvi. 3; Cant. ii. 5; Hosea iii. 1.

SHEMARIM is derived from *shamar*, to preserve, and has the general signification of things preserved. It occurs five times. In Exodus xii. 42, the same word, differently pointed, is twice translated as signifying *to be kept* (observed). Ps. lxxv. 8, dregs; Isa. xxv. 6, fat things; Jer. xlviii. 11, lees; Zeph. i. 12, lees.

MAMTAQQIM is derived from *mahthaq*, to suck, and denotes sweetness. It is applied to the mouth (Cant. v. 16) as full of sweet things. In Neh. viii. 10, "drink the sweet" *mamtaqqim*, sweetness, sweet drinks.

SHAKAR (sometimes written *shechar*, *shekar*) signifies sweet drink expressed from fruits other than the grape, and drunk in an unfermented or fermented state. It occurs in the Old Testament twenty-three times. Lev. x. 9; Numb. vi. 3 (twice wine and vinegar), xxviii. 7; Deut. xiv. 26, xxix. 6; Judges xiii. 4, 7, 14; I Sam. i. 15; Ps. lxix. 12; Prov. xx. 1, xxxi. 4, 6; Isa. v. 11, 22, xxiv. 9, xxviii. 7, xxix. 9, lvi. 12; Micah ii. 11. *Shakar* is uniformly translated strong drink in the Authorized Version, except in Numb. xxviii. 7 (strong wine), and in Ps. lxix. 12, where, instead of drinkers of *shakar*, the Authorized Version reads *drunkards*. (See "Generic Words.")

Greek, Latin, and English Generic Words

OINOS—Biblical scholars are agreed that in the Septuagint or Greek translation of the Old Testament and in the New Testament, the word *oinos* corresponds to the Hebrew word *yayin*. Stuart says: "In the New Testament we have *oinos*, which corresponds exactly to the Hebrew *yayin*."

As both *yayin* and *oinos* are generic words, they designate the juice of the grape in all its stages.

In the Latin we have the word *vinum*, which the lexicon gives as equivalent to *oinos* of the Greek, and is rendered by the English word wine, both being generic. Here, then, are four generic words, *yayin, oinos, vinum,* and *wine*, all expressing the same generic idea, as including all sorts and kinds of the juice of the grape. Wine is generic; just as are the words groceries, hardware, merchandise, fruit, grain, and other words.

Dr. Frederic R. Lees, of England, the author of several learned articles in *Kitto's Cyclopedia*, in which he shows an intimate acquaintance with the ancient languages, says: "In Hebrew, Chaldee, Greek, Syriac, Arabic, Latin, and English, the words for wine in all these languages are *originally*, and always, and *inclusively*, applied to the blood of the grape in its primitive and natural condition, as well, subsequently, as to that juice both boiled and fermented."*

*Used by permission of Sane Press, Oklahoma City, Oklahoma.

APPENDIX B

Correct Uses of the Hebrew Words Translated "Wine" in the King James Version Bible (in the Judgment of the Authors)

Genesis	9:21	*yayin*	Fermented Wine
	9:24	"	Fermented Wine
	14:18	"	Unfermented Wine
	19:32	"	Fermented Wine
	19:33	"	Fermented Wine
	19:34	"	Fermented Wine
	19:35	"	Fermented Wine
	27:25	"	Unfermented Wine
	27:28	*tirosh*	Unfermented Wine
	27:37	"	Unfermented Wine
	49:11	*yayin*	Unfermented Wine
	49:12	"	Unfermented Wine
Exodus	29:40	"	Unfermented Wine
Leviticus	10:9	"	Fermented Wine
	23:13	"	Unfermented Wine
Numbers	6:3	"	Fermented Wine
	6:3	"	Fermented Wine
	6:20	"	Unfermented Wine
	15:5	"	Unfermented Wine
	15:7	"	Unfermented Wine
	15:10	"	Unfermented Wine
	18:12	*tirosh*	Unfermented Wine
	28:7	*shekar*	Unfermented Wine
	28:14	*yayin*	Unfermented Wine

Deuteronomy	7:13	*tirosh*	Unfermented Wine
	11:14	"	Unfermented Wine
	12:17	"	Unfermented Wine
	14:23	"	Unfermented Wine
	14:26	*yayin*	Unfermented Wine
	16:13	*yeqeb*	Unfermented Wine
	18:4	*tirosh*	Unfermented Wine
	28:39	*yayin*	Unfermented Wine
	28:51	*tirosh*	Unfermented Wine
	29:6	*yayin*	Unfermented Wine
	32:33	"	Fermented Wine
	32:38	"	Fermented Wine
	33:28	*tirosh*	Unfermented Wine
Joshua	9:4	*yayin*	Fermented Wine
	9:13	"	Fermented Wine
Judges	9:13	*tirosh*	Unfermented Wine
	13:4	*yayin*	Fermented Wine
	13:7	"	Fermented Wine
	13:14	"	Fermented Wine
	19:19	"	Unfermented Wine
1 Samuel	1:14	"	Fermented Wine
	1:15	"	Fermented Wine
	1:24	"	Unfermented Wine
	10:3	"	Unfermented Wine
	16:20	"	Unfermented Wine
	25:18	"	Unfermented Wine
	25:37	"	Fermented Wine
2 Samuel	6:19	*ashishah*	Unfermented Wine
	13:28	*yayin*	Fermented Wine
	16:1	"	Unfermented Wine
	16:2	"	Unfermented Wine
2 Kings	18:32	*tirosh*	Unfermented Wine
1 Chronicles	9:29	*yayin*	Unfermented Wine
	12:40	"	Unfermented Wine
	16:3	*ashishah*	Unfermented Wine
	27:27	*yayin*	Unfermented Wine
2 Chronicles	2:10	"	Unfermented Wine
	2:15	"	Unfermented Wine
	11:11	"	Unfermented Wine
	31:5	*tirosh*	Unfermented Wine
	32:28	"	Unfermented Wine
Ezra	6:9	*chamar*	Unfermented Wine
	7:22	"	Unfermented Wine
Nehemiah	2:1	*yayin*	Fermented Wine
	2:1	"	Fermented Wine

Nehemiah (continued)

	5:11	*tirosh*	Unfermented Wine
	5:15	*yayin*	Unfermented Wine
	5:18	"	Unfermented Wine
	10:37	*tirosh*	Unfermented Wine
	10:39	"	Unfermented Wine
	13:5	"	Unfermented Wine
	13:12	"	Unfermented Wine
	13:15	*gath*	Unfermented Wine
	13:15	*yayin*	Unfermented Wine
Esther	1:7	"	Fermented Wine
	1:10	"	Fermented Wine
	5:6	"	Fermented Wine
	7:2	"	Fermented Wine
	7:7	"	Fermented Wine
	7:8	"	Fermented Wine
Job	1:13	"	Unfermented Wine
	1:18	"	Unfermented Wine
	32:19	"	Fermented Wine
Psalms	4:7	*tirosh*	Unfermented Wine
	60:3	*yayin*	Fermented Wine
	75:8	"	Fermented Wine
	78:65	"	Fermented Wine
	104:15	"	Unfermented Wine
Proverbs	3:10	*tirosh*	Unfermented Wine
	4:17	*yayin*	Fermented Wine
	9:2	"	Unfermented Wine
	9:5	"	Unfermented Wine
	20:1	"	Fermented Wine
	21:17	"	Fermented Wine
	23:20	"	Fermented Wine
	23:30	"	Fermented Wine
	23:30	*mimsak*	Fermented Wine
	23:31	*yayin*	Fermented Wine
	31:4	"	Fermented Wine
	31:6	"	Fermented Wine
Ecclesiastes	2:3	"	Fermented Wine
	9:7	"	Unfermented Wine
	10:19	"	Unfermented Wine
Song of Solomon	1:2	"	Unfermented Wine
	1:4	"	Unfermented Wine
	4:10	"	Unfermented Wine
	5:1	"	Unfermented Wine
	7:9	"	Unfermented Wine
	8:2	"	Unfermented Wine

Isaiah	1:22	*sobe*	Unfermented Wine
	5:11	*yayin*	Fermented Wine
	5:12	"	Fermented Wine
	5:22	"	Fermented Wine
	16:10	"	Unfermented Wine
	22:13	"	Fermented Wine
	24:7	*tirosh*	Unfermented Wine
	24:9	*yayin*	Fermented Wine
	24:11	"	Fermented Wine
	27:2	*chemer*	Unfermented Wine
	28:1	*yayin*	Fermented Wine
	28:7	"	Fermented Wine
	28:7	"	Fermented Wine
	29:9	"	Fermented Wine
	36:17	*tirosh*	Unfermented Wine
	49:26	*asis*	Unfermented Wine
	51:21	*yayin*	Fermented Wine
	55:1	"	Unfermented Wine
	56:12	"	Fermented Wine
	62:8	*tirosh*	Unfermented Wine
	65:8	"	Unfermented Wine
Jeremiah	13:12	*yayin*	Fermented Wine
	13:12	"	Fermented Wine
	23:9	"	Fermented Wine
	25:15	"	Fermented Wine
	31:12	*tirosh*	Unfermented Wine
	35:2	*yayin*	Unfermented Wine
	35:5	"	Unfermented Wine
	35:5	"	Unfermented Wine
	35:6	"	Unfermented Wine
	35:6	"	Unfermented Wine
	35:8	"	Unfermented Wine
	35:14	"	Unfermented Wine
	40:10	"	Unfermented Wine
	40:12	"	Unfermented Wine
	48:33	"	Unfermented Wine
	51:7	"	Fermented Wine
Lamentations	2:12	"	Unfermented Wine
Ezekiel	27:18	"	Unfermented Wine
	44:21	"	Fermented Wine
Daniel	1:5	"	Fermented Wine
	1:8	"	Fermented Wine
	1:16	"	Fermented Wine
	5:1	*chamar*	Fermented Wine
	5:2	"	Fermented Wine
	5:4	"	Fermented Wine

Daniel (continued)

	5:23	"	Fermented Wine
	10:3	*yayin*	Unfermented Wine
Hosea	2:8	*tirosh*	Unfermented Wine
	2:9	"	Unfermented Wine
	2:22	"	Unfermented Wine
	3:1	*ashishah*	Fermented Wine
	4:11	*yayin*	Fermented Wine
	4:11	*tirosh*	Unfermented Wine
	7:5	*yayin*	Fermented Wine
	7:14	*tirosh*	Unfermented Wine
	9:2	"	Unfermented Wine
	9:4	*yayin*	Unfermented Wine
	14:7	"	Unfermented Wine
Joel	1:5	"	Fermented Wine
	1:5	*asis*	Unfermented Wine
	1:10	*tirosh*	Unfermented Wine
	2:19	"	Unfermented Wine
	2:24	"	Unfermented Wine
	3:3	*yayin*	Fermented Wine
	3:18	*asis*	Unfermented Wine
Amos	2:8	*yayin*	Fermented Wine
	2:12	"	Fermented Wine
	5:11	"	Unfermented Wine
	6:6	"	Fermented Wine
	9:13	*asis*	Unfermented Wine
	9:14	*yayin*	Unfermented Wine
Micah	2:11	"	Fermented Wine
	6:15	*tirosh*	Unfermented Wine
	6:15	*yayin*	Unfermented Wine
Habakkuk	2:5	"	Fermented Wine
Zephaniah	1:13	"	Unfermented Wine
Haggai	1:11	*tirosh*	Unfermented Wine
	2:12	*yayin*	Unfermented Wine
Zechariah	9:15	"	Fermented Wine
	9:17	*tirosh*	Unfermented Wine
	10:7	*yayin*	Unfermented Wine
Matthew	9:17	*oinos*	Unfermented Wine
	9:17	"	Fermented Wine
	9:17	"	Unfermented Wine
Mark	2:22	"	Unfermented Wine
	2:22	"	Fermented Wine
	2:22	"	Fermented Wine
	2:22	"	Unfermented Wine
	15:23	"	Fermented Wine
Luke	1:15	"	Unfermented Wine

Luke (continued)

	5:37	"	Unfermented Wine
	5:37	"	Unfermented Wine
	5:38	"	Unfermented Wine
	5:39	"	Fermented Wine
	7:33	"	Unfermented Wine
	10:34	"	Not Certain
John	2:3	"	Not Certain
	2:3	"	Not Certain
	2:9	"	Unfermented Wine
	2:10	"	Unfermented Wine
	2:10	"	Unfermented Wine
	4:46	"	Unfermented Wine
Acts	2:13	*gleukos*	Unfermented Wine
Romans	14:21	*oinos*	Unfermented Wine
Ephesians	5:18	"	Fermented Wine
1 Timothy	3:3	*paroinos*	Fermented Wine
	3:8	*oinos*	Fermented Wine
	5:23	"	Not Certain, Probably Fermented
Titus	1:7	*paroinos*	Fermented Wine
	2:3	*oinos*	Fermented Wine
1 Peter	4:3	*oinophlugia*	Fermented Wine
Revelation	6:6	*oinos*	Unfermented Wine
	14:8	"	Fermented Wine
	14:10	"	Fermented Wine
	16:19	"	Fermented Wine
	17:2	"	Fermented Wine
	18:3	"	Fermented Wine
	18:13	"	Unfermented Wine

NOTES

Chapter 2

1. Robert L. Hammond, *Almost All You Ever Wanted to Know* About Alcohol *but didn't know who to ask!* (Lansing, Mich.: Michigan Alcohol and Drug Information Foundation, 1978), p. 30.
2. Ibid.
3. Ibid.
4. Ibid., p. 6.
5. Walter B. Knight. *Knight's Treasury of Illustrations* (Grand Rapids: Wm. B. Eerdmans Pub. Co., 1963), p. 390.
6. Ibid.
7. Hammond, *All You Wanted to Know*, p. 29.
8. Ibid., p. 33.
9. John Kobler, *Ardent Spirits* (New York: G. P. Putnam & Sons, 1973), p. 33.
10. Hammond, *All You Wanted to Know*, p. 8.
11. Ibid.
12. Knight, *Illustrations*, p. 394.

Chapter 3

1. Walter B. Knight, *Knight's Up-to-the-Minute Illustrations* (Chicago: Moody Press, 1974), p. 11.
2. Walter B. Knight, *Knight's Treasury of Illustrations* (Grand Rapids: Wm. B. Eerdmans Pub. Co., 1963), p. 397.
3. "Third Special Report to the U.S. Congress on Alcohol and

Health," from the Secretary of Health, Education, and Welfare (Rockville, Md.: National Institute on Alcohol Abuse and Alcoholism, June, 1978), p. 61.

4. Robert L. Hammond, *Almost All You Ever Wanted to Know* About Alcohol *but didn't know who to ask!* (Lansing, Mich.: Michigan Alcohol and Drug Information Foundation, 1978), p. 44.
5. Don Wharton, "What Two Drinks Will Do to Your Driving," *The Rotarian* (October, 1951).
6. Ibid.
7. "Report to Congress on Alcohol and Health," p. 63.
8. Ibid.
9. Ibid.
10. Ibid.
11. Billy Sunday, "The Trail of the Serpent," 1917 sermon, *The Sword of the Lord*, July 14, 1967.

Chapter 4

1. Arthur Fisher, "How Much Drinking Is Dangerous?" *New York Times Magazine*, May 18, 1975. Condensed in *Reader's Digest*.
2. Ibid.
3. Ibid.
4. Sidney Katz, "Booze: Why You Shouldn't Drink a Drop," Toronto *Star*, n.d.
5. Albert Z. Maisel, "Alcohol and Your Brain," *Reader's Digest* (June, 1970).
6. Robert L. Hammond, *Almost All You Ever Wanted to Know* About Alcohol *but didn't know who to ask!* (Lansing, Mich.: Michigan Alcohol and Drug Information Foundation, 1978), p. 27.
7. Ibid.
8. Ronald Kotulak, "The Chemical Sledgehammer: Alcohol Linked to Birth Defects," Chicago *Tribune*, September 13, 1977.
9. Walter B. Knight, *Knight's Up-to-the-Minute Illustrations* (Chicago: Moody Press, 1974), p. 70.
10. Ibid., p. 67.
11. "Third Special Report to the U.S. Congress on Alcohol and Health," from the Secretary of Health, Education, and Welfare (Rockville, Md.: National Institute on Alcohol Abuse and Alcoholism, June, 1978), p. 64.
12. Ibid.
13. "News from the World of Medicine," *Reader's Digest* (August, 1979).
14. Katz, "Booze," n.d.

Chapter 5

1. Roger Campbell, *Justice Through Restitution* (Milford, Mich.: Mott Media, 1977), p. 1.
2. Ramsey Clark, *Crime in America* (New York: Simon and Schuster, 1970), p. 30.
3. Clare Boothe Luce, "If Present Trends Continue, Democracy 'Is Bound to Collapse,' " *U.S. News and World Report* (July 5, 1976), pp. 65–66.
4. Ibid.
5. James Q. Wilson, *Thinking About Crime* (New York: Basic Books, 1975), pp. 43–44.
6. Walter B. Knight, *Knight's Treasury of Illustrations* (Grand Rapids: Wm. B. Eerdmans Pub. Co., 1963), p. 360.
7. Ibid.
8. Ibid., p. 390.
9. "Third Special Report to the U.S. Congress on Alcohol and Health," from the Secretary of Health, Education, and Welfare (Rockville, Md.: National Institute on Alcohol Abuse and Alcoholism, June, 1978), p. 64.
10. Tad Bartimus, "Alcoholism: Big Problem for Alaska," Chattanooga *News Free Press*, December 26, 1976.

Chapter 6

1. Robert L. Hammond, *Almost All You Ever Wanted to Know* About Alcohol *but didn't know who to ask!* (Lansing, Mich.: Michigan Alcohol and Drug Information Foundation, 1978), p. 50.
2. Ibid., p. 53.
3. *The Bottom Line*, Vol. 2, No. 4 (Winter, 1978), p. 7.
4. Hammond, *All You Wanted to Know*, pp. 54–55.
5. Ibid.
6. Ibid.
7. Walter B. Knight, *Knight's Treasury of Illustrations* (Grand Rapids: Wm. B. Eerdmans Pub. Co., 1963), p. 400. Used by permission.

Chapter 7

1. Norman H. Clark, *Deliver Us From Evil* (New York: W. W. Norton & Co., 1976), p. 22.
2. Donald Barr Chidsey, *On and Off the Wagon* (New York: Cowles Book Co., 1969), pp. 7–8.
3. Clark, *Deliver Us*, p. 4.
4. Ibid., p. 36.

5. Ibid., p. 39.
6. Walter B. Knight, *Knight's Master Book of New Illustrations* (Grand Rapids: Wm. B. Eerdmans Pub. Co., 1956), p. 671.
7. John R. Rice, ed., *The Best of Billy Sunday* (Murfreesboro, Tenn.: Sword of the Lord Pub., 1965), p. 76.
8. Ibid., p. 63.
9. John Kobler, *Ardent Spirits* (New York: G. P. Putnam & Sons, 1973), p. 12.

Chapter 8

1. Ross J. McLennan, *Booze, Bucks, Bamboozle & You* (Oklahoma City: Sane Press, 1978), p. 133.
2. Louise Bailey Burgess, *Alcohol and Your Health* (Los Angeles: Charles Pub., 1973), p. 152.
3. McLennan, *Booze*, p. 125.
4. J. C. Burnham, "New Perspectives on the Prohibition 'Experiment' of the 1920s," *Journal of Social History*, 2 (1968), p. 60.
5. As quoted in McLennan, *Booze*, pp. 142–143.
6. Norman H. Clark, *Deliver Us From Evil* (New York: W. W. Norton & Co., 1976), pp. 145–146.
7. Burnham, "New Perspectives on the Prohibition 'Experiment' of the 1920s," p. 63.
8. McLennan, *Booze*, pp. 118–120.
9. Burgess, *Your Health*, p. 152.
10. Ibid., p. 153.
11. Deets Pickett, *Then and Now* (Columbus, Oh.: School and College Service, 1952), p. 68.
12. Fletcher Dobyns, *The Amazing Story of Repeal* (Chicago and New York: Willett, Clark & Co., 1940), p. 381.
13. Ibid., pp. 387–388.
14. McLennan, *Booze*, p. 134.
15. Ibid., pp. 134–135.
16. Dobyns, *Repeal*, p. 375.
17. Ibid., p. 390.

Chapter 9

1. Mark A. Noll, "America's Battle Against the Bottle," *Christianity Today* (January 19, 1979).
2. F. B. Meyer, *The Five Books of Moses* (Grand Rapids: Zondervan, 1955), p. 16.
3. Matthew Henry, *A Complete Bible Commentary* (Chicago: Moody Press, n.d.), p. 17.
4. Joseph A. Seiss, *The Gospel in Leviticus* (Grand Rapids: Zondervan, n.d.), pp. 180–181.
5. Ibid., pp. 181–182.

6. William Patton, *Bible Wines or Laws of Fermentation and Wines of the Ancients* (Oklahoma City: Sane Press, n.d.), pp. 11–12.
7. Lloyd Button, "Did Jesus Turn Water Into Wine?" *The Baptist Bulletin* (February, 1973).
8. Patton, *Bible Wines*, p. 56.
9. Robert P. Teachout, "The Use of 'Wine' in the Old Testament," Th.D. dissertation, Dallas Theological Seminary, 1979, p. 312.
10. Patton, *Bible Wines*, p. 74.
11. Charles Wesley Ewing, *The Bible and Its Wines* (Denver: Prohibition National Committee, 1949), p. 10.

Chapter 10

1. G. H. Lang, *The Parabolic Teaching of Scripture* (Grand Rapids: Wm. B. Eerdmans Pub. Co., 1955), p. 45.
2. William Patton, *Bible Wines or Laws of Fermentation and Wines of the Ancients* (Oklahoma City: Sane Press, n.d.), p. 79.
3. H. A. Ironside, *Addresses on Luke* (Neptune, N.J.: Loizeaux Brothers, 1947), p. 177.
4. R. A. Torrey, *Difficulties in the Bible* (Chicago: Moody Press, 1907), pp. 96–97.
5. William L. Pettingill, *Bible Questions Answered* (Wheaton, Ill.: Van Kampen Press, n.d.), pp. 223–224.
6. Albert Barnes, *Barnes on the New Testament*, Luke-John volume (Grand Rapids: Baker Book House, 1949), p. 194.
7. Charles Wesley Ewing, *The Bible and Its Wines* (Denver: Prohibition National Committee, 1949), p. 17.
8. John R. Rice, *The King of the Jews* (Grand Rapids: Zondervan, 1955), p. 436.
9. Patton, *Bible Wines*, p. 85.

Chapter 11

1. H. A. Ironside, *Lectures on Acts* (Neptune, N.J.: Loizeaux Brothers, 1943), pp. 45–46.
2. William Patton, *Bible Wines or Laws of Fermentation and Wines of the Ancients* (Oklahoma City: Sane Press, n.d.), p. 93.
3. J. A. Seiss, *The Gospel in Leviticus* (Grand Rapids: Zondervan, n.d.), p. 183.
4. Walter B. Knight, *Three Thousand Illustrations for Christian Service* (Grand Rapids: Wm. B. Eerdmans Pub. Co., 1950), p. 670.
5. Robert P. Teachout, "The Use of 'Wine' in the Old Testament," p. 440.
6. Patton, *Bible Wines*, p. 112.
7. Charles Wesley Ewing, *The Bible and Its Wines* (Denver: Prohibition National Committee, 1949), p. 22.

Chapter 12

1. Walter B. Knight, *Knight's Treasury of Illustrations* (Grand Rapids: Wm. B. Eerdmans Pub. Co., 1963), p. 348.
2. Roger F. Campbell, *Let's Communicate* (Fort Washington, Penn.: Christian Literature Crusade, 1978), p. 171. Used by permission.
3. James R. Adair, *The Old Lighthouse* (Chicago: Pacific Garden Mission, 1966), p. 57.
4. Ibid., p. 59.
5. Walter B. Knight, *Knight's Master Book of New Illustrations* (Grand Rapids: Wm. B. Eerdmans Pub. Co., 1956), pp. 664–65.
6. Upton Sinclair, *The Cup of Fury* (Great Neck, N.Y.: Channel Press, 1956), p. 98.

Chapter 13

1. Walter B. Knight, *Knight's Treasury of Illustrations* (Grand Rapids: Wm. B. Eerdmans Pub. Co., 1963), p. 196.
2. "Facing Up to Alcoholism," pamphlet by the U.S. Dept. of Health, Education, and Welfare, Washington, D.C., p. 4. Printed in 1978.
3. From *The Trouble With Alcohol* by Tom Shipp, pp. 77–78. Copyright © 1978 by Eloise DeLois Shipp. Published by Fleming H. Revell Company. Used by permission.
4. Jack Odell, *Here's How!* (Grand Rapids: Zondervan, 1956), pp. 12–13.
5. Knight, *Treasury*, p. 400.
6. Odell, *Here's How!*, pp. 28–29.

Chapter 14

1. "Economic Strategies for Prevention," speech by William Plymat, executive director, American Council on Alcohol Problems.
2. Walter B. Knight, *Three Thousand Illustrations for Christian Service* (Grand Rapids: Wm. B. Eerdmans Pub. Co., 1947), p. 566.
3. George T. B. Davis, *When the Fire Fell* (Philadelphia: The Million Testaments Campaigns, 1945), p. 31.
4. Ibid., pp. 70–71.
5. *Alcohol Use and Alcoholism*, based on research by Dr. Wolfgang Schmidt and Mr. Jan de Lint (Addiction Research Foundation of Ontario).

INDEX